Managing the University: A Systems Approach

edited by
Paul W. Hamelman
foreword by
Linwood Holton
Governor of Virginia

The Praeger Special Studies program—utilizing the most modern and efficient book production techniques and a selective worldwide distribution network—makes available to the academic, government, and business communities significant, timely research in U.S. and international economic, social, and political development.

Managing the University: A Systems Approach

PRAEGER SPECIAL STUDIES IN U.S. ECONOMIC AND SOCIAL DEVELOPMENT

Praeger Publishers New York Washington London

PRAEGER PUBLISHERS
111 Fourth Avenue, New York, N.Y. 10003, U.S.A.
5, Cromwell Place, London S.W.7, England

Published in the United States of America in 1972
by Praeger Publishers, Inc.

Library of Congress Catalog Card Number: 75-186195

Printed in the United States of America

To G. J. H.

FOREWORD

The project on which this book is based first came to my attention when it was little more than a gleam in the eyes of its original supporters. At that time I was asked to release a small sum of money to help with the cost of presenting a symposium dealing with problems of management and planning in institutions of higher learning. This I was glad to do, and it is gratifying to see a continuing interest in this important subject by persons within and outside colleges and universities in the fifty states and abroad.

I am a lawyer by profession, not an educator or an economist. As governor, I am very much aware that persons with varying orientations and perspectives will look at the same problem quite differently. But too often equal attention is not given to those common points of agreement and basic propositions which can bring opposing viewpoints into closer harmony. This book is an attempt to highlight and elaborate generally accepted propositions while exploring applications of new concepts to the solution of increasingly complex problems in this vital area of our society. Two facts of national significance seem especially important for higher education at this time.

First, the nation's higher education system has just experienced a period of unprecedented growth and expansion. In no comparable period in our history has the rate of increase in public and private support for colleges and universities matched the record of the 1960's. Virtually all would agree that this has been a wise investment that will benefit the present and future generations. At the same time, however, evidence has appeared of imbalances that are too conspicuous to ignore. The low enrollment of minorities and low-income students in college in relation to their numbers in society as a whole and the oversupply of graduates in some fields in relation to national or regional demand are but two examples.

Second, our country will soon celebrate its 200th anniversary. In recognition of this occasion we can expect increased public awareness of national priorities, the meaning of freedom, and those governing principles that will continue to serve our democracy with equal effectiveness in its third century as they have in its first 200 years. One of the enduring definitions of the meaning of liberty is that of Justice Learned Hand: "The spirit of liberty is the spirit that is not too sure it is right." In some ways this definition accurately describes the spirit of academic freedom and scholarly inquiry. Without the freedom to seek and state the truth, institutions of higher learning will fail dismally in one of their most important functions in modern society. On the other hand, without the candor to recognize that yesterday's solution may no longer be valid for today's problem, the colleges and universities will also fail society. This book reflects a commitment to find new and better ways for institutions of higher learning to serve their many publics. It is important for the continuing vitality of our country and all its institutions that this spirit of inquiry be strengthened and preserved.

Linwood Holton
Governor of Virginia

PREFACE

In June, 1970, a symposium was held on the campus of Virginia Polytechnic Institute and State University with the theme "Systems Analysis for Higher Educational Planning." The objective of the conference was to explore the contributions from the fields of management science, operations research, organizational theory, and economics that could be used to help solve administrative problems which typically confront colleges and universities. A considerable interest in the program developed; and a substantial number of faculty, administrators, and state officials from Virginia and neighboring states—the audience for whom the program was designed—were represented at the conference.

Subsequent to the meeting a great deal of interest was expressed by members of the academic community in other states and overseas about the possibility of giving wider distribution to the ideas and papers presented at the conference. Accordingly, this book contains revised versions of the speakers' remarks.

The focus of this book has been well stated by Linwood Holton, governor of Virginia, who played a significant role in making the conference possible. Governor Holton's welcoming statement read in part:

> To me the appeal of the program is its concern with management. Management is not the principal ingredient of organization. Purpose certainly comes first, and resources are a necessity. Yet, both purpose and resources will become waste if management does not put one to the achievement of the other. Higher education organizations present special problems in administration. There is a delicate balance between the independence essential to advancing knowledge and the various constraints which require some direction. Intelligent innovation has characterized our academic pursuits. It is encouraging to see interest in the same approach to management which must provide the means.

Goals, priorities, the organizational imperfections of institutions of higher learning, the infinite array of attractive opportunities that a university could pursue but the finite set of resources available to achieve them—these are the recurring themes of the chapters in this volume.

How can systems analysis help improve the functioning of colleges and universities? The answer depends in large part on how one interprets the meaning of the term "systems analysis." To some the term conveys an image of computer-based information systems, massive data printouts, and detailed procedures to govern the behavior of individuals in an organization. Others would define systems analysis as an umbrella for a bagful of sophisticated techniques, including economic analysis, mathematical statistics, game theory, queuing theory, Delphi, and the like. Elaborate mathematical models and numerical calculations mark the work of those practitioners who view systems analysis as synonymous with a quantitative approach to problem solving. And still others consider the term to mean nothing more than a way of looking at problems of choice, although the approach requires a certain amount of quantitative common sense. By this definition the advantage of the systems approach is its aim of narrowing the discussion on substantive issues during controversies over goals, priorities, or allocation of scarce resources to competing objectives. The question of what is important becomes far more important than who is right.

These varying interpretations of the systems approach will be evident in the chapters of this book. One would not expect two systems analysts to approach the same problem in precisely the same manner, or necessarily to arrive at a common solution. It follows, therefore, that the contributors to this volume, only a few of whom would feel comfortable with the title "systems analyst," do not present a unified or internally consistent theory of the way institutions of higher learning are supposed to behave. Each contributor, by virtue of long association with academic institutions, is fully aware that colleges and universities are anything but trivial systems or organizations that are easy to understand.

And yet, for all their complexities, decisions must still be reached about the substantive questions of higher education planning—how much of what kinds of education, for whom, and at whose cost? H. A. Simon, commenting on the strictly managerial aspects of the educational enterprise, compares the role of university leader with that of executives in other fields of human endeavor in the following way:

> His organization's product—new and transplanted knowledge and skill—may seem a bit unusual when compared with the products of an automobile manufacturer, a purveyor of breakfast cereal, or the head of the municipal street department. But at base, the product is not very different from that of the research and development laboratory of DuPont, the U.S. Bureau

> of Standards, or NBC. In particular, the college is no more exempt than they from the pecuniary calculus, from the necessity of justifying its claims on society's resources.*

This volume is written in the spirit and belief that many of the innovative management concepts which have emerged in the last several years, ideas which the university itself has helped to foster and develop can be effectively employed in providing society the accountability that is due it.

August 16, 1971

---

*H. A. Simon, "The Job of College President," Educational Record, XLVIII, 1 (Winter, 1967), 68-69.

ACKNOWLEDGMENTS

This book is the end product of a project which spanned several months and directly involved a number of persons who gave generously of their time and attention. Throughout the project, I benefited enormously from ideas and suggestions put forward by my colleagues at Virginia Tech, as well as by persons affiliated with other universities and agencies in Virginia.

When the proposal to conduct the symposium on which this book is based was at a preliminary stage of discussion, T. Marshall Hahn, William E. Lavery, Leslie F. Malpass, and H. H. Mitchell were early and enthusiastic supporters of the idea. The members of the Research and Development Advisory Committee (RADAC) of the State Council of Higher Education for Virginia performed a dual function. Not only did RADAC agree to fund the project, but the committee also served as a much-needed sounding board during the planning stages of the project. Clifford L. Adams, Dennis Barnes, Warren Heemann, and J. Michael Mullen were especially helpful in this capacity. Special thanks are due to Robert L. Masden of the State Council for Higher Education for providing material assistance over the entire duration of the project and for demonstrating his uncanny knack for dissolving my problems rather than let me fret about how to solve them.

University campuses have traditionally provided the necessary climate and interpersonal "chemistry" that are important factors in research and scholarly inquiry. I consider myself fortunate to be a member of an academic community where these conditions have developed to an unusually high degree. Within this sphere of kindred and conflicting viewpoints I am especially indebted to Robert L. King, Edward Mazze, James F. Tucker, and Charles Schotta.

And, finally, the ladies—Martha Robinson and Linda Goodman—who typed the manuscript and promptly performed a large number of burdensome clerical chores.

CONTENTS

Chapter		Page
	Decision Makers	27
	Changes in Emphasis	27
	Conclusion	28
	Notes	28
4	OBSERVATIONS ON THE SCOPE OF HIGHER EDUCATION PLANNING IN THE UNITED STATES Charles A. Nelson	31
	Introduction	31
	Operating Systems	33
	Program Planning and Budgeting Systems	33
	Identification of Goals	34
	Definition of Objectives	34
	Program Description	34
	The Program Budget	34
	An Extended Time Frame	35
	Explicit Consideration of Alternatives	35
	Evaluation	35
	Replanning	36
	Management Information Systems	36
	Long-Range Plans	38
	Planning Models	42
	The State of the Art	42
	Conclusions	44
5	THE WICHE PLANNING AND MANAGEMENT SYSTEMS PROGRAM: ITS NATURE, SCOPE, AND LIMITATIONS Ben Lawrence	49
	Goals	49
	Guidelines for Program Development	50
	Areas for Research and Development in the Management Process	50
	Goal Setting	53
	Program Planning and Resource Allocation	53
	Execution	54
	Evaluation	55
	Communication Base	55
	Areas of Research and Development in Which WICHE Is Currently Working	56
	Relationship of WICHE-PMS Projects to the Higher Education Management Process	56
	Projects Concerned with the Development of Common Data Elements and Common Procedures	59

Chapter | Page

LIST OF TABLES

LIST OF CHARTS

LIST OF FIGURES

LIST OF ABBREVIATIONS

CAMPUS	Comprehensive Analytical Methods for Planning University Systems
CFP	Cost-Finding Principle
DED	Data Element Dictionary
DMS	Department Management Systems
FAA	Faculty Activity Analysis
FCF	Facilities Costs and Capital Financing
FTE	Full-Time Equivalent
HEFA	Higher Education Facilities Manual
HEFM	Higher Education Finance Manual
HENIS	Higher Education National Indicators Survey
IEP	Information Exchange Procedure
IO	Input-Output Indicators
ISDM	Information Systems Design Manual
MIS	Management Information System
PCM	Personnel Classification Manual
PCS	Program Classification Structure
PPB	Program Planning and Budgeting
PRIME	Planning Resources in Minnesota Education
RRPM	Resource Requirements Prediction Model
SAM	Space Analysis Manual
SCDP	Small College Demonstration Project
SCOPE	System for Control, Operations, Planning, and Evaluation

SFM	Student Flow Model
SWPMS	Statewide Planning and Management Systems
WICHE-PMS	Western Interstate Commission for Higher Education Planning and Management Systems (Project)

Managing the University: A Systems Approach

CHAPTER

1

A SYSTEMS VIEW OF UNIVERSITY ORGANIZATION

William E. Moran

It is appropriate that the first chapter of a book of this type be written by a person with diverse talents and experiences, both within and outside academia. William E. Moran is eminently well suited to the task by virtue of his past experience as a consultant to business and nonprofit organizations, as a university planner familiar with the promises and limitations of computer-based long-range planning models, and with the insights gained from his present position as president of a university. This chapter discusses some organizational aspects of colleges and universities, particularly the lack of a generally valid or clear-cut model of how universities actually function. Although institutions of higher learning are different from General Motors, the RAND Corporation, or a metropolitan police department, these differences in no sense exclude them from the public mandate for candid self-appraisal and management accountability. Ed.

This chapter reviews the possible contributions that a systems approach might make to the design of university organizations, improving institutional interrelationships and interactions with the external world. Characterizing the university as a system means simply that it consists of a set of interacting parts and exhibits some kind of integrity as a whole. Systems, whether biological, social, or mechanical, are ordinarily engaged in generating a product or achieving an objective. It is characteristic of a system that when one part is altered, the whole is altered in a greater or lesser degree through rearrangement of the parts.

A systems approach to organization design means simply trying to assess the whole organization rather than identifying one of the parts where difficulty may be most apparent. It means looking at interior interactions to understand why the organization is operating as it is, or to predict its behavior. It means studying the interaction of the system or organization with the sociopolitical, competitive environment within which it is located and with which it is inextricably linked. A systems approach concentrates upon relationships rather than parts.

ORGANIZATIONS IN DISARRAY

Perhaps the most striking feature of the last few years on college campuses in this country and abroad has been the extremes of disorder which have sporadically occurred. Disorder has not been universal. It has not even been destructive in all cases but occasionally has contributed to useful change. Further, hard-core criminal action, when it has occurred, is usually traceable to a relatively small number of persons. Nonetheless, one is struck by the apparent difficulty which the systems we call universities are experiencing in living with themselves, in expanding, in introducing change, and in preserving what should be preserved in terms of policy or practice.

The short-term ineffectiveness of many of our universities in dealing with change is obviously quite serious. It is serious because the country looks hopefully toward the intellectual resources of the university as a source of relief and guidance in handling the terribly complex social, political, economic, and, perhaps, spiritual problems of the day. It has discouraged many to find that our bastions of reason are filled with people who are all too human, all too susceptible to emotional upset, to bias, to self-interest, and to fear of change. But if a common reaction to the last few years has been one of disappointment or frustration, it must be remembered that satisfaction ordinarily is a function of expectations.

Too much has been expected of our campuses, and in that sense we are responsible for our own disappointment. Prevailing models of what universities are and should be have been sentimental and archaic. Chester Barnard, formerly president of the Rockefeller Foundation and president of the New Jersey Bell Telephone Company, observed in 1938:

> It is readily believed that organized effort is normally successful, that failure of organization is abnormal . . . but in fact, successful cooperation in or by formal organizations is the abnormal, not the normal condition . . . what are observed from day-to-day are the successful survivors among innumerable failures.

> . . . Failure to cooperate, failure of cooperation, failure of organization, disorganization, disintegration, destruction of organization—and reorganization—are characteristic facts of human history.[1]

A smoothly functioning university is neither an easy nor a permanent achievement. It is not to be taken for granted. It is not in the natural order of things for universities to survive over many years without a great deal of stress and chaos felt internally and at the boundary between the organizations and the environments of which they are a part. If it is correct to believe that university organizations are, like other formal organizations, obliged to maintain themselves internally and to adapt themselves to rapidly changing environments, it is instructive to consider the structural characteristics of universities and compare them with those of other organizations.

An organization chart for a modern university presents the familiar network of president, vice-president, deans, and so on arranged in what appears to be classical hierarchy. That is, a vice-president for finance with supporting staff appears to be in charge of financial and business problems. An academic vice-president is inevitably to be found on the chart, and he appears to be responsible for academic problems of all kinds. Other functional specialists may also be found in public relations, research, legal affairs, student services, and so on. This hierarchical arrangement of functional specialists is, however, only one of the existing decision structures. There are others in the wings which are brought onstage by certain kinds of decision requirements. The existence of other decision structures made up of interrelated powerful committees, senate bodies, and student groups, cutting across hierarchical lines, is a recognized phenomenon of university life. These different structures, as we know, tend to bump into each other.[2]

Why do student groups, faculty committees or senates, and administrators bump into each other so? Why does each tend to see the other as overstepping its "proper" bounds, as exceeding its jurisdictional rights? The answer may well be that the reality of the university as a social system consisting of interacting hierarchical and communal decision structures has never really been understood. One wing of most universities is convinced that "town meetings," rather than administrators, should govern the campus. Another likes to think that flat-out bureaucracy is essential for campus peace and order.

THE UNIVERSITY AS A BALANCED SYSTEM

A systems approach does not accept the idea that either process of decision making (communal or hierarchical) can successfully ignore the other, or that one approach to governance must dominate others. What is more, a systems view is uncomfortable with the assumption that nonacademic administration can be easily separated from academic decision making. Many persons honestly believe that universities contain faculty and academic subunits which can be—indeed, should be—wholly removed from administrative affairs, budgeting, or facility planning with which hired help is paid to deal. This view of university organizations is the antithesis of a systems view. It is also a long way from the truth, for "nonacademic" problems have the most irritating way of cutting across organizational structure and even entering the classroom if unchecked. Issues recur regularly which are not narrowly financial or narrowly academic or narrowly public relations. As a result of this resistance to simple classification, problems which must be dealt with by universities prove to require close working relationships between the vice-president for finance and the academic vice-president, as well as between faculty, administration, and students. Between, in other words, community and hierarchy.

By viewing the campus organization as a system of delicately balanced hierarchical and communal relationships, one is likely to be less optimistic that it can be repaired by replacing an ineffective component, as one would insert a new spark plug into a sputtering car. A new president for a campus which has been racked with disorder and filled with anger is not necessarily a solution, even if the campus is lucky enough to get a good man. The graduation or dismissal of a militant student leader may not help the situation very much. They are attractive choices with all the earmarks of a solution for those who like simplistic models. It is certainly true that if the formal leader of an organization is replaced, the organization, whether a university or a commercial corporation, may be affected significantly. If, however, the problems that are present within the university concern the relationship of the components with one another, the newly elected president will find these relationships remain to be repaired when he arrives.

An effective university president is often described as one who refuses to accept "bureaucracy" on his campus, who insists upon making a community of the university. When Weber coined the term and described the bureaucratic model, he included in his definition a number of characteristics which are common in universities: regulations, departments, hierarchy, schedules, and so on. Lest one imagine that it is only the fiscal side of universities that operates bureaucratically, he should recall the ranking of professors and their characteristic work specialization. Bureaucracy is clearly there. Weber had some reservations about what bureaucracies could become, but he imagined a highly rational, dispassionate organization model. What is more, he concluded that bureaucracy is virtually unavoidable, that the choice in organizing human effort is between dilettantism and bureaucracy. The latter had at least the potential of efficiency and effectiveness in the ordering of human affairs.

University organizations are complicated social systems. They contain bureaucracies but are not merely bureaucracies. Other decision networks are present: committees and constituencies which exercise great influence on university behavior and decision making. The notion that effective institutions commonly rely upon "participative democracy" to the exclusion of bureaucracy is false. It is equally mistaken to deny the communal nature of a university and to try to place in the hands of university administrators the kind of power which accrues to military or corporate leaders. Affiliation with a university is not the same as affiliation with a federal agency or General Motors. University organizations are actually fragile combinations of bureaucratic and communal decision structures. These two entities interpenetrate and are inseparable. The relationship is essentially symbiotic.

THE USES OF MULTIPLE STRUCTURE

One of the chief advantages of bureaucratic organization is that it provides a decision structure which is capable of responding rapidly in a preplanned way to certain kinds of problems. The disadvantages in practice are too well-known to require elaboration; but recent intensified pressure upon university organisations to respond to social problems, to account more adequately for their resources, to plan their own development, and in general to manage their affairs seems to require the presence of bureaucracy.

Bureaucracy as a model or in practice is not enough. Both the teaching faculty and the student body, whose personal development is at the heart of university purpose, must be permitted to speak on questions affecting present and future goals of the university and to shape the processes by which the university solves its problems. An effective university organization enjoys a fairly sympathetic, trustful working relationship between bureaucracy, faculty, and student constituencies. In such an organization the bureaucracy will respond to problems which are most effectively and expeditiously handled by the bureaucratic decision structure while the faculty and the students, or both in combination, will respond to other questions which affect the vital interests of both and for which both can speak with experience and responsibility. Other matters will prove to be ineffectively handled unless bureaucracy, students, and faculty respond jointly and cooperatively. Any other approach to designing the internal relationships of a university usually devolves to the analogy of three scorpions in a bottle, each gingerly circling the others, awaiting opportunity to strike the decisive blow. Instances of such institutional madness are not rare.

The difficulty in campus decision making is simply that on one occasion the university is obliged to respond with the precision and maneuverability of a Panzer division while on another the appropriate decision process may be a meeting of faculty and students not unlike a New England town meeting. It is possible for organizations to shift from one structure to another. In a modern university it is not essential for one of the structures ultimately to dominate the other. What is crucial is that the decision rules by which a university shifts from one decision structure to another—say, from hierarchy to faculty senate—should be well understood and agreed upon by most members of the organization. This is very close to the heart of the matter, and it is not simple to arrange.

If a multiple structure organization is not only desirable but also essential to the survival of universities, what should be the criteria by which one structure will be selected over another in the face of serious problems? No general answer can be given. The question is all the more thorny when the inflow rate of problems shifts suddenly or when the significance of a familiar issue mushrooms unpredictably. New issues do not easily fit static decision rules or static organization. Competition among multiple structures for the authority to respond to a given situation can be worse for a university than no response at all. Destructive competition of this sort can shake a campus to its very foundations.

The ad hoc faculty-student crisis committee is a telltale sign of ineffective institutional decision making. One social scientist has labeled the crisis committee a "synthetic structure." It often appears in the face of major community disasters. It is common in university organizations under stress. J. D. Thompson concedes that synthetic structures can accomplish some useful ends, but adds:

> . . . they are not efficient . . . the overriding reason is that the synthetic organization must simultaneously establish its structure and carry on operations . . . it must assemble and interrelate the components and it must do all of this without benefit of established rules or commonly known channels of communication.[3]

If the fact finding of the Cox Committee at Columbia (reporting on the disturbances there) is to be believed, the evidence supports Thompson:

> One of the minor lessons to be drawn is that constant talk and numerous intermediaries, both authorized and

> volunteer, are more likely to be a source of confusion than assistance.[4]

Of the Ad Hoc Faculty Group, a voluntary mediating body at Columbia, the Cox Committee goes on to say:

> . . . the Ad Hoc Faculty Group's role seems also to have suffered from inherent defects . . . a group can never mediate effectively. The negotiated adjustment of a crisis requires clean lines of communication, central direction, and, on the part of the mediator, the ability to impart and receive confidential communications.[5]

Multiple-structure decision processes cannot be devised in the midst of fearful difficulty. Nor can the question of criteria be considered coolly during a strike or sit-in. While it is probably impossible to produce a set of rules which will remove all uncertainty about how a university should respond, a start must be made in peacetime. Some degree of uncertainty always accompanies multiple structure. It follows that a degree of trust is absolutely essential to effective campus decision making. It is necessary on occasion, when time is short, for hierarchy to act unilaterally on matters that affect vital campus interests. Such action must be optional, since the objective of organization design is response to real problems under real constraints. A shortage of response time may be one of the constraints. When action is taken that is inconsistent with perceived decision rules, however, it must be done with good cause. Such action should be followed by a thorough discussion between the leaders of the hierarchical organization and leaders of student or faculty groups whose legitimate options were usurped.

Without a high degree of trust between constituent parts of a university, shifting of decision structures predictably yields suspicion and fear. These comments have alluded only to faculty, students, and administrators. The potential complexity of multiple structure goes beyond a faculty senate or student polity working in conjunction with administrators. There may be literally dozens of significant constituencies present simultaneously in the system.

The concept of program budgeting which has attracted so much attention in the last few years really implies an interesting new version of multiple structure for universities. Imagine for a moment the kinds of goodwill and trust required between departmental chairmen and program administrators. Matrix organization is implicit in program budgeting, and much lubrication is needed where the lines of authority cross. That lubrication is intrainstitutional trust. Multiple structure per se cannot guarantee good results. An organization with multiple decision structures which are neither discussed nor subject to review and change, which unintentionally overlap one another at crucial times and in critical areas, offers very little promise for the future. Further, in the absence of effective resource and academic program planning, the pressure upon university decision structures is simply too great to be borne. The operating problems rooted in poor planning can overwhelm any decision network, multiple or single.

Pat answers for fitting together a university's internal components should be viewed with suspicion. It is characteristic of open systems that they are capable of solving problems in more than one way. There are many ways to structure the bureaucratic elements present on campus and also many ways in which students, faculty, and hierarchy might relate to one another. This does not mean that all alternative organizational designs are equally good. Further, a design that has proved effective at one point in time may be ineffective a few years later. Anticipating the need to change systemic relationships pays great dividends.

THE CAMPUS AND ITS ENVIRONMENT

Assessing the whole of an organization requires more than a comprehension of decision structures. External relationships of the campus must be traced as well. Living systems work to maintain themselves internally while achieving their purposes and adapting to the environments of which they are parts. A living system is always "open," or interacting with what is outside itself. Each campus draws its members from beyond its organizational boundaries. It requests and receives resources from alumni and/or legislative bodies and must answer to these external supporters in order to survive. To be open, therefore, is to be vulnerable.

Some campuses have suffered in the past from a delusion consisting of two main parts. The first is that the dozens of millions of dollars required to operate universities or build physical facilities are provided because faculty hearts are pure and their purposes noble. The second is that social change flows only one way. We like to believe that we are agents of social change, that innovation in society should be developed conceptually at our institutions and then extended into the political and social systems of the land. But it is unhappily true that we support the cause of change in the abstract and assume that it is other people, other groups, other organizations—rather than our own—which must change. The liberal view of universities with regard to social change and our conservatism in terms of university overhaul is a matter of record.

A systems approach to planning the campus takes legislative and alumni bodies, and even townspeople, seriously. It means keeping them reasonably informed about campus plans and operations and even sharing the planning process. It means anticipating the kinds of questions and misunderstandings that may exist over the next five-ten years and dealing with them in advance, on the assumption that what happens in the environment really does count.

Some very important steps have been taken in explaining institutional objectives and accounting for university resources. Development of management information systems, standardized accounting procedures, space utilization data, and comprehensive personnel information is taking place. This has come about largely through probes from the outside rather than through anticipation of information requirements by campus leaders—but it is happening. Much has been done in the last few years in these directions. Full disclosure is still a long way off.

A remaining issue is the matter of accounting for faculty time. Contact hours and some student advising are formally planned and scheduled. A faculty member's activities in research, departmental administration, public service, and a host of other activities are less well understood even on campus. An outsider might expect that any inquiry into the many-faceted responsibilities of teaching faculty would be welcomed as an opportunity to demonstrate performance. In actual fact there is often a good deal of resistance to attempts to track faculty through the academic year and to understand better how faculty time is actually being expended. There are good and bad reasons for such resistance. Some faculty members feel that the freedom which has always been associated with independent research may be curtailed in some undesirable way. Mistrust of the potential uses to which such data could be put presents a very real obstacle in attempting to gather such sensitive information. These data are essential, however, for planning a university and justifying its operating budgets, since teaching faculty represents a very large percentage of operating costs. If and when this is done, it will be incumbent upon those collecting and storing information to see that the purposes for which it has been gathered are not quietly revised. The trust of teaching faculty should not be abused. It would be quite unfair for administrators to track faculty activity, ostensibly for budgetary or planning purposes, and then to use that information formally or informally for purposes of evaluation and promotion.

The time has come for public and private institutions to develop a more generous approach toward sharing information which discloses organizational objectives, plans, and resource utilization data with faculty, staff, and students as well as with relevant publics off campus. The paradoxical inference to be drawn from the history of the last decade is that universities are intrinsically fragile social systems which must nonetheless be opened to the public for both inspection and use.

It is certainly true that open systems do not survive for very long if they are not selectively open. A campus which is not careful about strings attached to gifts, the kinds of students admitted, the quality of faculty, and the caliber of administrators has no future at all. Similarly, political pressures to teach in a particular way or to evaluate faculty by their ideological leanings should not be permitted to be a part of any university evaluation process. That facet of the academic environment must be kept out at all costs.

Like the individual who was surprised and delighted to learn that he had been speaking prose all his life, experienced administrators and faculty are likely to be partially familiar with what has been said here. This partial familiarity, rooted in experience rather than systems theory, is an organizational asset that can be very helpful in planning an institution's future. It is vitally important to understand that public and private institutions can no longer live in splendid isolation, removed from public problems and unconcerned with the neighborhoods in which they are located. Precisely how a given institution should relate itself constructively to the sociopolitical environment of which it is a part remains a complex matter. A systems approach does not lead to simple answers. It does provide some assurance that the right questions will be asked.

NOTES

1. Chester I. Barnard, The Functions of the Executive (Cambridge, Mass.: Harvard University Press, 1938), pp. 4-5.

2. W. E. Moran,"The Study of University Organizations," Journal of Higher Education, XXXIX, 3 (March, 1968), 144-155.

3. J. D. Thompson, Organization in Action (New York: McGraw-Hill, 1967), pp. 52-53.

4. Crisis at Columbia: Report of the Fact Finding Commission (New York: Vintage Books, 1968), p. 146.

5. Ibid, p. 155.

CHAPTER

2

SOME REFLECTIONS ON HIGHER EDUCATION ADMINISTRATION

Martin L. Shotzberger

What current forces might be expected to exert a significant influence on the scope and character of higher education in the 1980's? How well do abstract models of the decision-making process merge with the organizational imperfections of colleges and universities? As power becomes more diffuse among many campus constituencies, what tools does the university leader have at his disposal to influence the out-comes of the decisions that are made, or to reconcile the honest differences of opinion that surround many campus issues? These and related questions are the focus of this chapter. Ed.

Administrators in higher education have several things in common with administrators of other types of institutions: (1) their positions require knowledge and skills with which few were originally endowed; (2) all of them are required to manage—in the conceptual sense of the term; (3) many of them move into college administration because they have been effective performers in other areas which brought to light their accomplishments, rather than having been trained, educated, or prepared to fill the academic position; and (4) most are left to learn the job by doing it—a dreadful experience for those who believe that knowledge is a prerequisite for professional performance.

There are many current forces and developing trends which will affect higher education in the immediate, intermediate, and long-range future. Since these facts and trends are so well-known, I shall not cite them but, rather, emphasize the basic overall impact they are likely to have on the management of institutions of higher learning. It seems that we can expect the following:

1. An increasing demand for more education—a possible increase of more than 100 percent in head count from 1967 to 1977, and this may be low[1]

2. An increasing demand for better education, as measured in quality of all educational resources[2]

3. An increasing insistence upon a greater variety of higher educational opportunities—technical, humanistic, community, research, graduate, and continuing human development, for example[3]

4. An insistence upon efficiency and effectiveness, with particular reference to effective utilization of resources

5. A reduction of a monopoly on research

6. A reduction in administrative autocracy and/or authority

7. A greater demand that the administrator lead managerially, rather than administer in the narrow sense of the word.

The last two items will serve as the springboard—perhaps a paradoxical one—for what will be the substantive remarks of this chapter.

MANAGEMENT STYLES IN TRANSITION

It may appear difficult to us to lead if we are to have a reduction in authority. Difficult as it sounds, it is probably true. The basic reason for reduction in authority results from what one must consider the rapidly developing freedom of our basic educational resource—the faculty—plus the freedom and sophistication of the student. As the component parts of higher education become more free, knowledgeable, and mobile, the more they desire influence in the decisions which affect them. Thus the style of the administrator must change from a position of authority to one of professional colleague. Yet he is in a position of central importance, and he wields great influence in the allocation of resources. Even though his position is powerful, his resources are scarce and the competition for them is great.

In the not distant past, the basic premises of administrative leadership seemed to be (1) a leader sets policy because of his position; (2) a leader orders rather than persuades; (3) only leaders are highly motivated; (4) there should be no change or innovation in established procedures; (5) one does not need knowledge if he has enough authority and power.

Today's leader, in institutions of higher learning and elsewhere, is seen quite differently. The leader influences policy. His tools of influence are persuasion, communications, and the human relations approach. He assumes high motive, integrity, and honesty on the part of others. He exhibits the innovative spirit. He uses knowledge as the base for his decisions and is fully aware that he is not the sole source of knowledge in the organization.

How can we manage our institutions of higher learning better? This question becomes especially basic if we accept the premises of a reduction in authority and an increasing demand for leadership. I should like to propose five basic responses: (1) know what administration or management involves, (2) become keenly aware of and knowledgeable about decision theory and its application, (3) develop an awareness of systems analysis, (4) develop an understanding and knowledge of communications theory and its application, and (5) develop a knowledge framework of human relations.

Of these five possible areas of development only the first three will be considered here.

First, in regard to knowing what management or administration is, it is helpful to model the process visually or schematically rather than by words alone. It is often said that management is "getting things done through people." This is too simplistic and does not suggest what or how.

The following general model (see Chart 1) I have found to be of great value in understanding and teaching management, as well as in practice. This is a generic model and varies only in degree at levels of administration.

From this model one would infer that it does not explain the kind, quantity, or quality of knowledge an administrator needs in order to function properly. Surely it is necessary to know information theory, decision theory, planning theory, organizing theory, control theory, and the theory of objectives.

In the late 1940's and for some time after, the purveyors of computer hardware inadvertently did management a disservice by suggesting that systems analysis would permit us to manage by exception, that is, to make routine all the repetitive tasks in an organization and bring to the attention of top management only those problems for which no solution had been devised. To manage by exception is not a bad approach, but it is not entirely adequate in today's complex organizations. I suggest that we should manage by objectives, as follows: (1) by routine objectives in large numbers, (2) by exception as little as possible, and (3) by innovation to the greatest extent of our released resources which result from sound routine. The effective

CHART 1

A Management Model

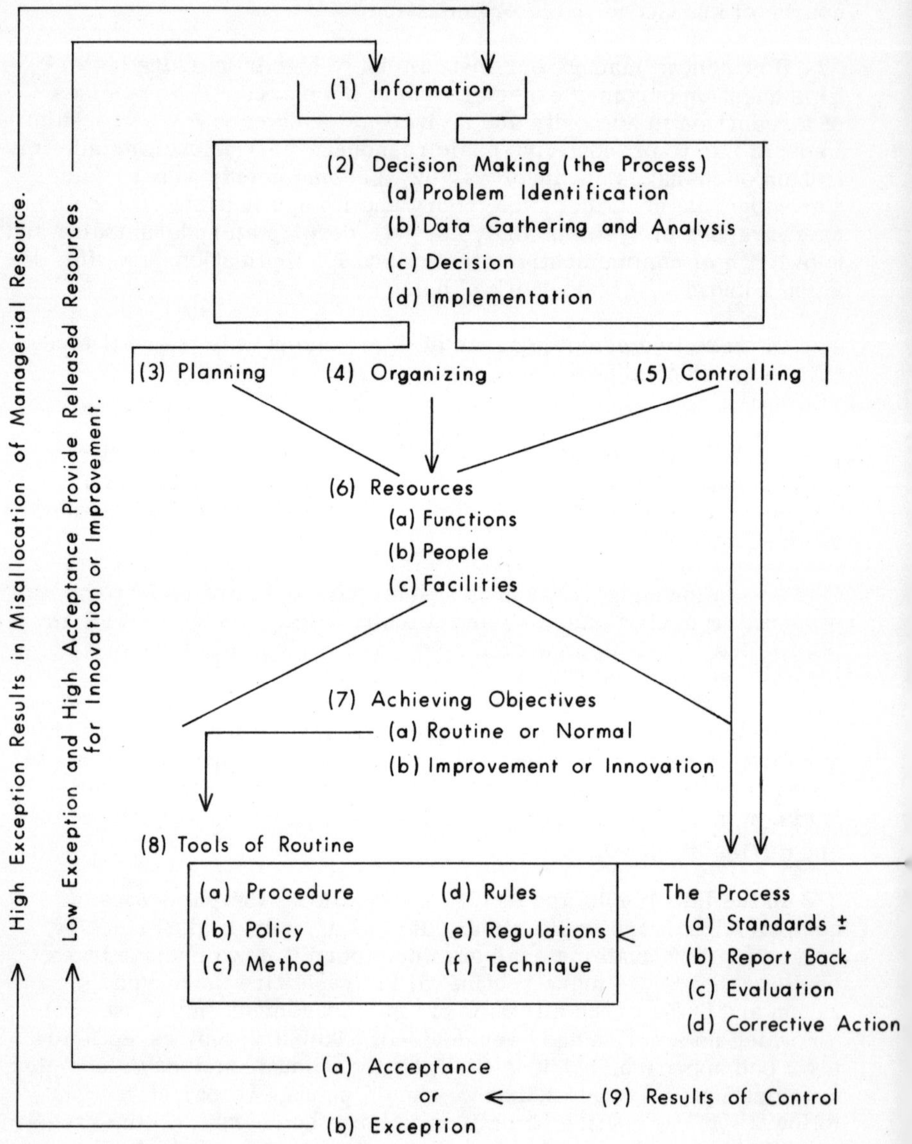

manager must, however, be extremely careful that routine does not become an obsession.

DECISION THEORY

Administrators are usually quite adept at establishing control processes and procedures, so I shall not dwell on this subject. It is important, however, to pursue some thought on decision theory.

There is a distinction between decision making and the decision-making process. The latter term is a process, as indicated in the model, whereas a decision involves arriving at a choice from among an array of alternatives or possibilities. Arriving at a decision, descriptively, is not much more than mental games, and of itself is rather sterile. Therefore the remainder of the managerial process is essential. It is necessary at this point to suggest that a manager is, in the decision-making process, singularly important only in arriving at the decision. For problem identification, data gathering, and analysis, as well as implementation, the manager must rely on resources other than himself, within and outside the organization.

Decisions have as their function the causing of a response within the organization. They are made under conditions of certainty, uncertainty, and risk. Conditions of certainty suggest that the outcome is almost totally assured; payroll and certain scheduling situations are examples. Conditions of risk imply a large number of decisions—so large a number that reasonable calculations may be made in the aggregate but not in the specific. Insurance coverage is a fair example. Conditions of uncertainty are those for which we have difficulty predicting the outcome because we know little in terms of our data base and/or the response to decisions. What we seek is high certainty in routine decisions and predictability in innovative ones, without the hazard and expense of large numbers of decisions. In the latter case we seek the predictability of the aggregate in the specific. How can this be done? The basic and general answer to the question depends on how one defines rationality.

There are three critical measures of successful decision making: quality, speed, and consistency.

Quality, of course, is measured by the achievement of stipulated objectives. Speed must be measured in absolute and relative time. Consistency measures the proportion of times that objectives are satisfactorily achieved.

What is rationality? Rather than thinking of it in an "either/or" context, it should be placed on a continuum from totally rational to totally irrational.

Purely rational decisions exist when the following conditions exist:

1. All information is available

2. A complete codifying ability is available

3. There is a full ability to foresee all possible choices

4. There is a full ability to evaluate all possible choices

5. There is a full ability to foresee all consequences

6. There is full availability and control over all necessary resources.

All of these will assure a theoretically perfect decision. It is rarely, if ever, that such exists.

When we have none of these conditions, we have total irrationality. There are many choices, however, between these extremes. We can evaluate some of these choices against the stipulated criteria of success, as shown below.

CRITERIA OF DECISION

	Speed	Consistency	Quality
Pure Rationality	Absolute—slow Relative—fast	High	High
Deductive (if the principle[s] is [are] sound)	Fast	High	High
Inductive (if, in the research sense, the criteria are met as in rationality)	Slow	High	High
Policy and Procedure	Fast	High	High
Value Premises	Fast	High	Questionable (especially

	Speed	Consistency	Quality
			if all value premises are not accepted within the organization)
A Priori Analysis (if the assumptions are truly sound)	Fast	High	High
(other conditions)	Fast	Low	Low
Instuition	Usually fast	Low	Low

(Intuition is viewed as a form of induction without research, which may result in an unsound generalization that may serve as a decision base.)

Emotion	Fast	Low	Low

The question is often asked, "What percentage of good administrative or managerial decision is acceptable?" It is impossible to give a precise answer, but certainly it has to be extremely high; perhaps nearly 100 percent for routine decisions and in excess of 90 percent for nonroutine. Failure to make good innovative decisions results in extremely poor allocation of resources—a result no college or university can afford. Most of us do not do this well. Why? Probably because we incline toward the less rational tools of decision making. This results in large part from our failure to stipulate our objectives in percise terms, and as a result we fail to seek the proper information.

For my own purposes, I have found the following decision model to be of great assistance both when I was a dean and in my present position. (See Chart 2.) Awareness of such a model tends to keep one from moving to the low-order decision tools of value premises, intuition, and emotion.

The procedural and policy decisions are the most rational decisions we make, assuming they have been well planned and programmed. We often hear great administrators described as being highly intuitive. I believe you will find that they are really great deductive thinkers who have a vast store of knowledge and who are able to "spin their tapes," cut out the irrelevant, and find the point. They perhaps have the intellectual equivalent of the assumptions of pure rationality.

CHART 2

A Model of Managerial Decision Making

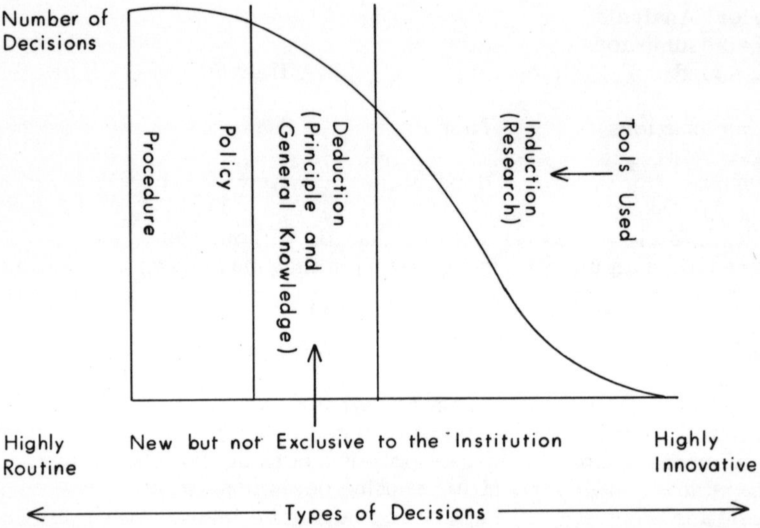

THE SYSTEMS VIEWPOINT

A system may be viewed from two extreme viewpoints: the specific system and the global system. In general, the best application of a specific system is that of a relatively simple mechanical device. An example of a global system may be society as a whole. Systems model builders very often want to ascribe to the global the characteristics of the specific while attempting to ascribe to the specific the characteristics of the global. The one approach is a false type of induction, and the other a false form of deduction. Both omit or add a great deal which, on sensible analysis, would not hold.

What is a system? In its simplest sense it is a sequence of inputs which are processed and result in outputs. The jargon, then, is inputs-processor-outputs. In the early days of systems analysis it was thought that the inputs were presented, then processed through some mysterious device which handled, manipulated, and dealt with these inputs and miraculously presented us with outputs. The processor and its work were called a "black box" (of which little was known) and were deemed to be truly mysterious. This rather absurd analysis gave us great difficulty because it led us to the conclusion that the only way we could get a better output was to vary inputs on a trial-and-error basis until just the right output emerged. While we in colleges and universities did not use this jargon, our behavior suggests that we operated this way. Although our "inputs," i.e., students, changed markedly in terms of aspirations and aptitudes, our "processor" changed hardly at all. Consider the stagnant curricula, the principle of in loco parentis, tenure, selection procedures, and such which, until recent times, served as our basis for academic behavior. Our behavior administratively was probably worse, in that we simply refused to acknowledge the variation in our inputs because we did not know what the processor would do. Perhaps, too, we did not want our output to vary, although, by and large, it did vary because of new disciplines.

Fortunately, through rather startling developments in operations research, computer applications, and other technologies, we now know the "black box" is not a mystery and is susceptible to description, analysis, study, control, and, most of all, understanding and use.

As administrators we have seen the extremely valuable use of systems analysis in the routine of our institutions, such as payroll, registration, library records, and the like, as well as in research and, to some extent, teaching. It is time for us to learn to use this approach in our innovative practices. To do so, however, we must (1) understand the basic principles in the systems approach, (2) become aware of the critical tools of systems, (3) be aware of the shortcomings and pitfalls in systems, and (4) able to specify what we need to know and why, particularly in regard to our objectives.

CONCLUDING REMARKS

I have already suggested the jargon of input-processor-output, but another important aspect of systems theory is the use of models. Models may be verbal, quantitative, or both—but in any case they are models, not pure reality. Even though not totally real, they can encapsulate and simplify extremely complex processes into understandable terms. I attempted to do this in a management model and in a decision model. Such a practice is possible for communications and human relations. The process is applicable in different areas in different ways. A map of Virginia, a set of blueprints for building, and a state plan for higher education are all models. I have seen multimillion-dollar firms modeled organizationally on a single page, both in process and in function. The Polaris missile program was modeled, as were all our space flights. Given this state of the art, who can say a college or university cannot be modeled?

The major pitfall of a system is that of entropy. A mechnical system, for example, will wear out if not repaired constantly, in that the inputs, the process, and the outputs are so fixed. Fortunately, colleges and universities are not mechnical systems. Although we use mechanical processes and procedures, our basic input is human—and it is a viable input. Good, viable human inputs can prevent entropy in our institutions. The administrator has great influence over the performance of these inputs and can, in his own right, preclude entropy in his organization.

We need to learn much about managing by objectives. This means more than our important, but not specific, statement of purpose. We should be stating objectives in terms of quantities and qualities of outputs, at what costs, with what resources, and by what devices. We need to know how to allocate resources within our system and how to construct research designs in general terms.

If we can do all this, we can likely make good use of technical staff people in systems which will result in rational decisions with high relative speed, high quality, and great consistency. We will, moreover, be able to implement the management process more effectively.

It is time that college and university administrators learn to use the new knowledge of the universe. We must learn the management process, the rational decision processes, and systems theory. We must become administrative theorists, not administrative hypothesizers. Leave the hypotheses to the researchers. They have time, talent, and other resources to test them. The administrator does not.

NOTES

1. Alvin C. Eurich, ed., Campus 1980 (New York: Delacorte Press, 1968).

2. John Caffrey, ed., The Future Academic Community (Washington, D.C.: American Council on Education, 1969).

3. Charles G. Dobbins and Calvin B. T. Lee, eds., Whose Goals for American Higher Education? (Washington, D.C.: American Council on Education, 1968).

CHAPTER

3

ORGANIZATIONAL CONSIDERATIONS IN EDUCATIONAL PLANNING

James F. Tucker

In general, university leaders have put primary emphasis on strengthening the basic academic departments on campus and have been reluctant to divert scarce resources to indirect services and educational support activities—less productive "overhead" items in the college budget. This ordering of priorities is commendable, but it has created some difficulties: a profusion of budgetary units at the departmental level and the absence of a rapid and consistent decision-making structure on campus. Whether it is a multiversity or a small college, a vast number of small decisions end up on the president's desk, with no one willing or able to take charge during his temporary absences from campus. Professor Tucker suggests some possible remedies for this situation in this chapter. Readers who are interested in the possible role faculty members might assume in this matter should refer to McGeorge Bundy, "Faculty Power," Atlantic Monthly, CCXXII (September, 1968), 41-7. Ed.

Many organizational considerations in higher educational planning are not related solely to the economic efficiency of the institution's expenditures or to the equity of the cost and benefits of the services dispensed but, rather, to providing the institution with a capacity to meet the priorities of a rapidly changing and complex society. This assumes that an educational system which produces the highest net economic returns and the most rapid economic growth may not be one which promotes an equitable society.[1] With this in mind, planners assessing the returns from expenditures on higher education should realize that many of the benefits, as well as the cost of higher education, do not accrue directly to the student. Indirectly, the student's family, government, and specific sectors of the community reap certain advantages, not all of which are economic in nature.

NEW MISSION AND SCOPE

Like many other factors in our rapidly changing society, the traditional mission and scope of our senior colleges and universities are undergoing a transition that will no doubt have serious implications for those involved in planning for higher education. Whereas the traditional mission and scope placed great emphasis on providing students with the opportunity to advance their intellectual and creative capacities, to develop students as responsible citizens, and to furnish them with the basic tools of analysis for becoming a part of the trained manpower required by a complex society, the focus now is on educating decision makers, conducting research, and proposing solutions to social problems in the hope that our colleges and universities can have a considerable, if not decisive, influence on the decisions that affect the nation's quality of life.[2]

Failure by secondary schools to achieve reasonable success with all of their clientele; pressures on the home and family to facilitate certain changes in attitude and behavior considered at variance with our present mores; the absence, in the majority of communities, of cohesive activity for the most active age group in our society; the existence of severe and swift repercussions in the average job situation for violations of well-defined regulations and expected behavior; and apathy on the part of too many eligible voters have forced the modern college and university to serve as an experimental laboratory for society. In the case of the secondary schools which have surrendered to the problems of low achievers and the families which have recoiled at the thought of boyfriend visiting daughter's bedroom until two o'clock in the morning, the answer has been, "Let the college and university do it." With regard to the communities, the young people recognize the social ills while they are still living in the community, but the recognition in large groups can best be facilitated on the college campus.

Another change in the mission and scope of higher education which is more related to the formal educational process, but still new to the college and university function, pertains to high school-level "catch-up" courses. Up until a very few years ago these offerings, usually in English and mathematics, were confined to the predominantly Negro institutions. They were often taken without credit and were designed to help these institutions boost their enrollments because they could not draw enough students who could meet the regular entrance requirements.

Recently, however, this business has grown tremendously. Not only have many predominantly non-Negro institutions found it necessary to adopt such courses in order to promote further integration of their student bodies, but a number of larger universities with recently adopted open admission policies have found it necessary to establish distinct organizational units to administer an entire program of remedial courses. These universities have discovered that it takes courses in subjects besides English and mathematics to bridge the gap between the high school preparation received by these youths and the kind of preparation needed to cope with the first year of study in a genuinely baccalaureate program. As a result, a number of senior colleges and universities have either already established or are considering the establishment of a special unit to administer a foundation curriculum.

Another key area which has been added to the mission and scope of colleges and universities involves more meaningful and direct service to those groups in the population which do not now derive any benefit from higher education. Many persons in these groups contribute handsomely to the budget, by way of taxes, for higher education in their states and local communities. Often they do not have children who attend college and, more often than not, the cultural and other extracurricular activities which their institutions open to the general public are not aimed at the noncollege-educated audience.

Of course, many of our colleges and universities can point to programs which they have conducted for that part of the population which is not actively seeking a college education. These programs, aside from extension activities, would include adult education, reading classes for adults, courses in arts and crafts, and other classes closely associated with the industrial arts. Despite these efforts, however, there are still many statesmen, civic leaders, students, and even faculty members who feel that colleges and universities are failing to respond to the community. ("Community" used in this sense is really to be interpreted as an "urban" area.) These persons feel that very few institutions of higher education have accepted the real world of the city on their doorstep as a laboratory in which they can advance certain intellectual pursuits.[3] As a result, colleges and universities are under tremendous pressure to undertake such ventures as setting up medical clinics and legal assistance offices in low-income neighborhoods, providing training courses for ghetto leaders, rendering technical assistance to the municipal government, and examining their own employment practices as well as those of their contractors with respect to equal opportunity.

All of the above developments are relatively new to the mission and scope of higher education and must be accommodated in the institutional organization because society has so directed. And, while there is still no assurance that such developments will persist indefinitely and make for permanent changes in the college and university organization, there is strong reason to believe that their presence will be felt until genuine progress is made toward curing some of the most glaring of our social ills.

ACADEMIC UNITS

One of the most delicate problems facing both the chief administrator and the planning official in the college or university relates to the administrative unit within the academic division of the institution. This problem would be simplified if all subject matter departments employed the same number of teachers, enrolled the same number of students, occupied the same amount of space, and used the same mode of operation in dispensing their services. However, such is not the case. Thus, key academic administrators and the planning officials must arrive at some understanding on the criteria for identifying academic units in order to facilitate an equitable allocation of the institution's resources.

As faculty members continue to stress the narrow specializations within subject matter disciplines, the trend will continue for the establishment of an increasing number of subject matter departments along with bureaus, institutes, and other structural devices, with no minimum staff size or level of operations. This in turn means that budgets for such items as supplies, equipment, travel, contractual services, recruiting, student wages, and scholarship funds sometimes must be broken down into amounts too small for effective utilization. There is also the prospect of having an academic division top-heavy with administrative and overhead costs and more department heads than the level of operation warrants.

Even with the larger academic units such as schools, colleges, divisions, or branches, where there is little question about the need for separate budgets for items such as supplies, equipment, and travel, consideration must be given to the need for separate catalogs, news-letters, registration, commencement exercises, financial record-keeping, and motor pools. Decisions on these matters depend a great deal on the organizational structure of the institution; and the mere fact that a unit is assigned the title of "school," "college," or "division" does not necessarily mean that it functions just like a unit with the same title at another institution. In fact, it may not function in the same manner as another unit with the same title at the same institution.

DECISION MAKERS

A combination of the growing need for instant decisions on the campus and the increasing number of off-campus appearances required of key administrative officials points up the need for providing each of these officials with an assistant or some other person with a title designating a "back-up" function. As an organizational consideration of planning, such a structure would not only cushion the institution against untimely delays in the decision-making process from the lowest level of administration up to the top but would also assist in assuring prompt consideration and implementation of the directives from the planning function itself.

Many of the grievances leveled against the bureaucracy of the college and university recently have pinpointed the slowness with which the institution moves in responding to a request for consideration on even the most routine item of operation. In fact, academic institutions generally have an image which is depicted as an easy-going, time-consuming, and "don't-rush-anything" operation. Basically, however, this image is misleading. The problem, in most cases, is that the only official empowered to give a definitive response to a request is engaged off the campus or away on his well-deserved one-month vacation. Whereas it may be a bit cumbersome to provide each decision maker with a "second-in-command," some consideration should be given to an organizational structure which could cushion the institution against untimely delays in the decision-making process from the lowest level of administration up to the top.

Planning officials, as well as key administrators in the college and university, ought to be extremely sensitive to an organizational mechanism for assuring prompt decisions in the institution. In some instances this may be accomplished by splitting the time of one person over two functions, with one carrying essentially a "back-up" responsibility. The other alternative, of course, is to build a second slot into a key position, with the clear understanding that one of the two officials will always be available.

CHANGES IN EMPHASIS

In response to students' demands for more autonomy in non-academic affairs and the concern of our courts of law for the guarantee of due process for students in college disciplinary proceedings, there is little question that some rearrangement will be necessary in the student affairs areas of our colleges and universities.

Generally, this rearrangement must be sensitive to the more recent posture of the student, in which he asserts himself more as a citizen and less as a subject for the institution's in loco parentis role. The housing and dining hall sections in this new structure will no doubt experience a decrease in staff, activity, and responsibility as the unit assumes a purely management or liaison function with private owners of student housing units and food services. At the same time, there will be an expansion in those sections which deal with such issues as the confidentiality of information about students, disciplinary proceedings, responsibility for the student when he or she is engaged in off-campus activities, the landlord-tenant relationship between the institution and the student, and coordination of on-campus and off-campus security forces. From an overall point of view, it may become increasingly difficult for the typical dean of students organization to serve as adviser to students and monitor student organizations while also serving as the prosecutor in disciplinary matters.

CONCLUSION

Contrary to a number of opinions which have been expressed so frequently and loudly during the past two or three years of campus unrest, most educational administrators have recognized the need for change in our institutions of higher education. Moreover, these same administrators have had the vision to foresee that many of these changes would involve adjustments in the institution's organization, and that such adjustments should be supported by careful planning which treats the individual college or university as an integrated system.

At the moment, no one can predict the nature of the next crisis in higher education, nor any of the organizational considerations that might be required to resolve the issues involved to any degree of satisfaction. However, there is reason to believe that close cooperation between key administrators and planning officials can keep higher education sufficiently dynamic to meet the challenge and still retain both the solvency and the integrity of our colleges and universities.

NOTES

1. "Introduction," The Economics and Financing of Higher Education in the United States, a compendium of papers submitted to the Joint Economic Committee of the United States, October 7, 1969, Ninety-First Congress, First Session (Washington: U.S. Government Printing Office, 1970), p. 3.

2. Campus Tensions: Analysis and Recommendations, report of the Special Committee on Campus Tensions (Washington: American Council on Education, 1970) p. 36.

3. John W. Gardner, "The University and the Cities," Educational Record, L, 1 (Winter, 1969), 5.

CHAPTER

4

OBSERVATIONS ON THE SCOPE OF HIGHER EDUCATION PLANNING IN THE UNITED STATES

Charles A. Nelson

This chapter marks a transition in the volume—from a consideration of strictly organizational aspects of colleges and universities to the discussion of the kinds of information needed to improve their functioning. Charles A. Nelson describes the interdependency of operating systems, program planning and budgeting systems, management information systems, long-range planning, and planning models. Among the many cogent ideas presented by Mr. Nelson, one is particularly intriguing: that the creation of a management information system does not preclude the possibility of participative management in higher education but, on the contrary, does much to promote it. Ed.

INTRODUCTION

American universities—and most of the universities of the world as well—are going through a time of troubles. They are enduring an avalanche of criticism from every quarter: from students, faculty members, trustees, legislators, the general public, even from university administrators. Many university leaders are incredulous that an institution which has undergone such extensive change in the last two decades can stand accused of being unresponsive to the needs of the day. Has the American university, while appearing to adjust, not really changed after all? Has it changed, but for the worse? And now that it is under attack, how should it respond? by standing firm? by new initiatives? And is planning universities for change as relevant for the American institutions as for their sister institutions around the world?

How do higher education planners respond to these new conditions? The impression is that most of those engaged in university planning—and here I refer to the professional planning officers, as distinct from the presidents and chancellors—are trying to ignore the current unrest. They seem to feel that the disruption is temporary and the campus will soon return to normal. However, the current crisis is raising such fundamental questions about the nature and operations of the university that if the planners ignore the current events, they will soon be planning for a university that no longer exists. The persistent questions that are being raised include the following: Who is entitled to a place in the university? Is the present faculty suited to the needs of the new students? Are our campuses getting too big? What is the appropriate role of students and of faculty in the governing structure? Who has the primary responsibility for maintaining order on the campus? Is the faculty paying enough attention to its teaching responsibilities? What sort of research is it appropriate for the university to undertake? In the familiar jargon of the planners these questions concern admissions policy, faculty qualifications, optimum size, organization structure, teaching loads, and research policy. These questions form the essence of the planning process in higher education.

The evidence is all around us that our universities are not adequately prepared for change—not prepared to anticipate it or foster it—indeed, not even prepared to combat it. The planning process is one in which organizations decide what they will seek and determine how to obtain it, clinging to that which is good in present policies and practices but just as assiduously replacing outworn goals and methods with new ones. The planner, then, must be acutely aware of the necessity for change. More important, the planning systems, procedures, and categories which are devised must accommodate change, not frustrate it.

Effective planning thus calls attention to the importance of gathering information on specified matters to serve as critical indexes of success. In other words, a good information system must provide the data necessary for evaluation. In addition, it should serve the needs of control, operations, and planning. For planning, information is needed to help determine what should be done. For operations, information is needed in order to do it. For control, information is needed in order to determine whether it is being done as planned. For evaluation, information is needed to determine whether the planned outcome has been achieved.

Before describing some aspects of the work done with respect to the management information needs of American universities, it is helpful to define some of the crucial terms.

OPERATING SYSTEMS

The term "operating system" refers to a combination of methods, procedures, equipment, and personnel organized to accomplish a specific repetitive function. Within the university, examples of operating systems would include the following:

1. The process of enrolling students in courses and the maintenance of student records

2. The processing of bills for payment

3. The maintenance of physical plant facilities and the scheduling of personnel required for such maintenance

4. The scheduling of classroom assignments (space allocation).

An operating system is designed to perform a specific task; the major tests of its efficacy are the accuracy, speed, and economy with which the job is accomplished. On any given university campus, it is quite possible that one operating system may be conducted with supreme efficiency while another is in disarray. For example, a university may do a superb job of maintaining its student records but perform very poorly with respect to the maintenance of its physical facilities; such a condition it is not uncommon. The point of the remark is that operating systems are not always interdependent. (It should be mentioned also that they are not equally dependent upon the computer for student records to be computerized and inefficient and for a physical plant maintenance system not to be computerized but highly efficient.)

Substantial improvements have been made in the efficiency of operating systems at universities. However, they frequently are not designed to provide the kind of information needed by senior administrators responsible for planning and evaluation.

PROGRAM PLANNING AND BUDGETING SYSTEMS

A program planning and budgeting (PPB) system is a comprehensive planning process that includes a program budget as its major component. A PPB system has a number of characteristics that sharply distinguish it from other, less sophisticated approaches to planning and budgeting.

Identification of Goals

Any educational effort should begin with some explicit identification of a social or individual need to which the system must address itself. At this level of generality, such a goal may be simply "to help produce a better-educated populace" or "to prepare young adults for gainful employment." It would hardly be necessary to mention this elementary step in the planning process if it were not for the fact that so many organized activities continue to function in our society because of the flow of inertia even after the need for them has disappeared.

Definition of Objectives

Defining objectives carries the process well beyond initial identification of goals. It calls for the specifying of the goal in a quantity over a given period of time. Thus: "to graduate fifty lawyers in 1969, of whom 95 percent will pass the bar examination within one year" or "to equip 200 classrooms with thirty student stations each by the opening of the fall term of 1972." It is hardly an exaggeration to say that while most universities can state their goals, few have explicit objectives of this sort.

Program Description

A program is a group of interrelated activities organized to accomplish an objective. In the case of a proposed new school of architecture for a university, for example, the program description would include a statement of the relation of the parts of the curriculum to one another, and perhaps the course outlines. It would also specify the requirements with respect to students, faculty, support services, and facilities.

The Program Budget

The program budget, usually covering a one-year period, states the expected costs in relation to the proposed programs instead of relating costs to objects or functions. Thus the cost of carrying out a program to accomplish an objective must be calculated. The program budget becomes a statement of policy, since it reveals the decisions as to how the available financial resources are to be applied to the accomplishment of stated objectives by means of specified programs. And since resources are almost invariably in short supply, the program budget reveals the choices that have been made among competing objectives and programs.

An Extended Time Frame

An extended time frame requires the planner to show what demands the proposed program will generate over a period of years—perhaps five or more. How much will it tax the available resources: money, facilities, and personnel? How much will it produce in return? This element in the process has grown out of the unhappy experience of many government agencies in discovering that the approval of a given project, which could readily be financed during the first year or two of its operation, creates unexpected strains upon resources five years later. In universities it is especially important that research programs be subject to such scrutiny, for they often entail commitments that extend beyond the initial funding period, and often they do not carry their share of overhead costs.

Explicit Consideration of Alternatives

Explicit consideration of alternatives is one of the most powerful and peculiar characteristics of the PPB system in operation. The advocate of a given program must specify the other possible approaches that could be taken to achieve the stated objective and must show in orderly fashion that the plan offered is superior to the available alternatives. For example, a state higher education coordination board might consider an alternative to a proposed new graduate program in social work to be the reliance upon another university for the preparation of graduates under a contractual arrangement. Or if a program calls for the expansion of a given campus from, say, 5,000 students to 10,000 students, an alternative would be to create one or more new campuses instead of expanding the old. The PPB system imposes the discipline of requiring that the alternatives be examined and a convincing case be made for the preferred program.

Evaluation

Evaluation is a two-part process. First, upon submission of a new program or continuation of an old program, the submitting party states the criteria by which the program is to be measured. At the completion of the specified period the criteria are applied and the success of the program is evaluated. Perhaps the most powerful impact of this process is that no activity, no matter how firmly established, is exempt from regular critique. There is no automatic funding of programs simply because they are entrenched.

Replanning

Replanning completes the PPB cycle and reintroduces into the system those modifications which result from the processes of evaluation. As appropriate, programs are dropped, revised, or expanded and new programs are devised to meet newly perceived goals. The planning process is thus seen as a continuous cycle of activity.

With these definitions and descriptions of operating systems and PPB systems in mind, the nature and characteristics of the management information systems can be readily introduced.

MANAGEMENT INFORMATION SYSTEMS

By a management information system (MIS) in higher education we mean that configuration of men, machines, and methods which supports management in the collection, storage, processing, and transmission of information for operation, control, evaluation, and planning of a university. Seen in this way, the MIS draws upon the operating systems and supports both the PPB system and the administrative management and control system of the institution. An important distinction now emerges: various operating systems may perform independently of one another, but the MIS is entirely dependent upon the adequacy of the various operating systems for the successful performance of its functions. Thus it appears that in the process of developing an MIS, the university will typically find that substantial revisions in its operating systems are required if the needs of the MIS are to be met.

The relationship of the MIS to the operating systems and to the PPB system may be seen in Chart 3.

The value of the MIS depends upon the informativeness and timeliness of its outputs, which in turn depend heavily upon the quality of the operating systems. It monitors the operating systems, providing selected information needed by management for control, and extracts from the operating systems and from specified external data (especially economic and demographic) information required for planning and evaluation. The outputs of an MIS consist ideally of the optimum combination of regular reports and ready responses to specific interrogations.

It is apparent, then, that the decisions as to what information is appropriate, in what form, for whom, and with what frequency are the critical elements in the design. It is important to stress that the

CHART 3

System for Control, Operation, Planning, and Evaluation in Higher Education

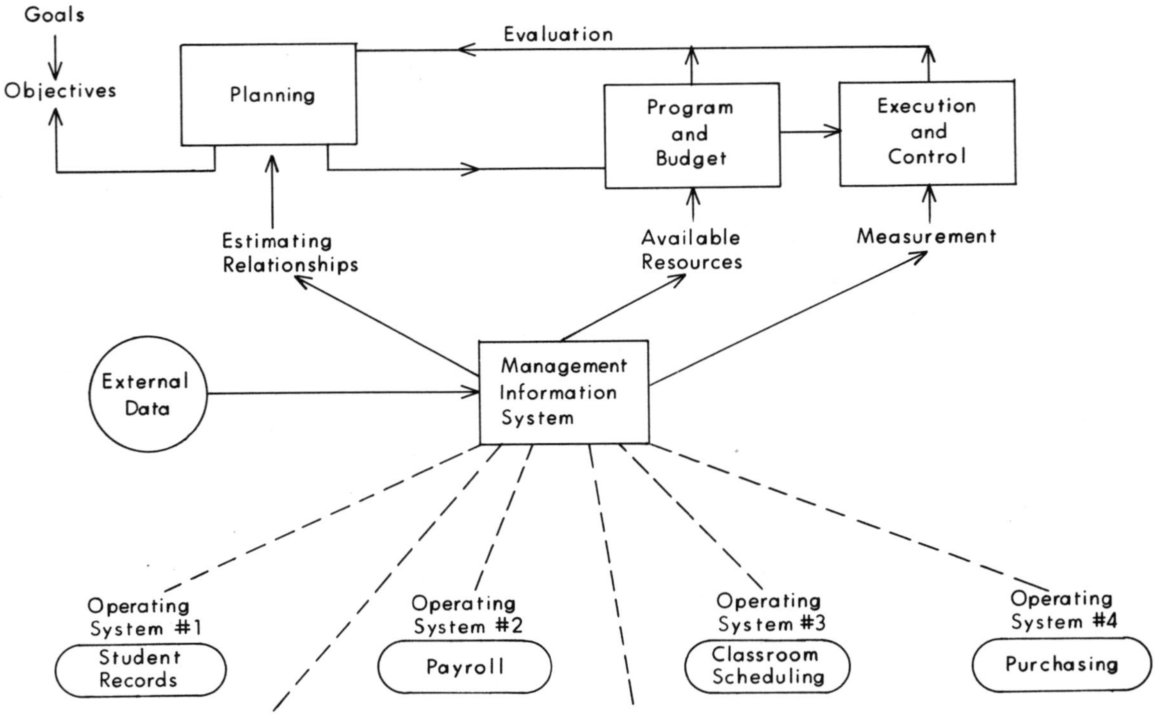

utility of such a system is not dependent upon a centralized, hierarchical organization structure. In many American universities authority is parceled out among trustees, the president, the faculty, and even the students. The president is not always the most powerful figure on the campus. Since participative management is clearly on the increase, it should be noted that information systems are even more important to its success than to the success of an organization in which all authority stems from a single source. The reason for this is that dependence on individual recollection of data and intuitive response to situations is sharply reduced as decision making is more broadly shared. A common body of information available to each decision-making group to meet its special needs becomes essential.

Very little about computers has been said thus far, but it will be apparent that for a complex institution such as a university, the computer is the only efficient means for accomplishing the desired results in a timely fashion. The output will be computer-produced and the inputs to the MIS will derive largely from computerized operating systems.

In an MIS recently designed for a state university the regular outputs to top administrators consist of forty-seven reports. Of these, twenty are planning reports and twenty-seven concern control and evaluation of operations. The reports in their entirety cover the academic program, students, faculty, support services, facilities, and finances. In the operating reports administrators are assisted in monitoring current operations by means of critical indicators and by comparison of actual results with budgeted figures. The planning reports provide updated annual projections for five years and a tenth-year trend projection as well.

LONG-RANGE PLANS

The long-range plan projects demands and requirements over a span of years. It differs from the expanded time frame of the PPB system in at least two respects: it spans a longer period of time; and it is developed at a more general level—that is, the program elements are dealt with in summary fashion rather than in the detail called for by the PPB procedures. Such a plan is particularly important for public institutions or agencies having the responsibility to provide programs, teachers, facilities, and funds for an entire generation of college students. The long-range plan may be incorporated into the planning function of the MIS. Table 1 provides an example of long-range planning data for a hypothetical university.

TABLE 1

Alpha State University Planning Summary

Division:_______________

Academic Year: Date:

	Current Commitment		Budget Year	Planning Year						
	Last Year		This							
Item	Planned	Actual	Year	1	2	3	4	5	10	15
A. Program Data:										
Instruction										
1. Total no. of departments										
2. Total no. of programs by level										
3. Total no. of majors										
4. Total no. of courses										
5. Total no. of sections										
6. Total enrollment (head count)										
7. FTE enrollment										
8. Average class size										
9. Total faculty (head count)										
10. Total faculty FTE										
11. Student/faculty ratio										
12. Faculty load (credit hours taught)										
13. Total student credit hours offered										
14. No. of degrees earned (by level)										
15. No. of financial aid recipients (by level)										
16. Faculty salaries										
17. Faculty fringe benefits										
18. Average faculty salary (FTE faculty)										
19. Average faculty fringe benefits (FTE faculty)										
Research										
1. No. of projects										
2. Total research dollars (by fund source)										
Public Service										
1. No. of short courses										
2. No. of conferences, workshops, institutes										
3. No. of community development programs										
4. No. of professional refresher courses										
Staff:										
1. Professional or administrative (FTE)										
2. Clerical (FTE)										
3. Other (FTE)										
4. Student employees (FTE)										
5. Total staff salaries and wages										
6. Total staff benefits										
7. Average staff salaries (FTE)										

Table 1 (Cont'd)

Division:______________

	Academic Year:							Date:			
	Current Commitment Last Year		Budget Year This Year	Planning Year							
Item	Planned	Actual		1	2	3	4	5	10	15	
B. STAFF: cont'd											
a. Professional or administrative											
b. Clerical											
c. Other											
d. Student											
8. Average staff benefits (FTE)											
a. Professional or administrative											
b. Clerical											
c. Other											
d. Student											
C. Endowments and Quasi Endowments:											
1. Amount (market value)											
2. Amount (book value)											
3. Yield on market value											
D. Gifts:											
1. Current purposes											
2. Endowment											
3. Plant and equipment											
4. Other											
E. Plant and Equipment:											
1. Book value											
2. Additions to physical plant											
3. New space requirements											
Income											
F. Educational and General:											
1. Tuition and fees											
2. Governmental appropriations											
3. Endowment income											
4. Gifts											
5. Sponsored research (grants)											
6. Other sponsored research (contract)											
7. Recovery of indirect costs											
8. Organized activities											
9. Other											
Subtotal											
G. Student Aid:											
1. Gifts											
2. Endowment											
3. Governmental appropriations											
4. Other											
Subtotal											

Division: ______________

Academic Year: Date:

Item	Current Commitment Last Year Planned	Current Commitment Last Year Actual	Budget Year This Year	Planning Year 1	2	3	4	5	10	15
Income cont'd										
H. Auxiliary Enterprises:										
1. Intercollegiate athletics										
2. Residence halls										
3. Food service										
4. Student union										
5. Other										
Subtotal										
Total Income										
Expenditures										
I. Educational and General:										
1. Instruction and dept. research										
2. Organized activities										
3. Sponsored research										
4. Other sponsored programs										
5. Public service										
6. Libraries										
7. Student services										
8. Plant operation and maintenance										
9. General administration										
10. General institutional expense										
Subtotal										
J. Student Aid:										
1. Scholarships										
2. Fellowships and prizes										
3. Tuition remission										
Subtotal										
K. Auxiliary Enterprises:										
1. Intercollegiate athletics										
2. Residence halls										
3. Food services										
4. Student union										
5. Other										
Subtotal										
Total Expenditures										

PLANNING MODELS

Recent articles and papers appearing in management science journals and educational literature reveal a substantial interest in computer-based planning models. It appears that the application receiving the greatest attention is enrollment projection. However, the development of a simulation of the entire planning process appears to be feasible. The development of planning models for specific universities, encompassing the full range of university activities, appears to be imminent.

The great virtue of the simulation model is its ability to reveal quickly the consequences throughout the university structure of any proposed change in conditions or policies—whether it be a reduction in enrollment, a rise in faculty salaries, a change in investment policies, a shift in the student/faculty ratio, an increase in research contracts, or any of many other possibilities. The speed with which the computer calculates and reports the consequences of such changes enables the planning staff to consider so many more possible courses of action than heretofore that it revolutionizes the planning process. In fact, it appears that the simulation model is the only practical way of considering the full implications of more than one or two alternative courses of action when a complex decision is involved.

THE STATE OF THE ART

Discussions of the state of the art are often unsatisfactory because of the very different interests and perspectives of the participants. What art are we talking about? Higher education planning? Computer applications to university operations? PPB? MIS? Model building?

By means of the concept of a system for control, operations, planning, and evaluation (SCOPE), we have attempted to provide a framework in which all the elements of this broad discourse can be accommodated. In effect, the MIS, PPB, and operating systems all become subsystems within SCOPE. Within this frame-work we may then discuss the state of the art as a whole and in its parts. The observations which follow are based on conversations, reading the literature, work with university clients, and efforts to learn what others are doing. In addition, it is unfortunately true that some institutions appear to be reporting their aspirations along with their achievements. Before each statement, assume the words "So far as we know . . . ."

A fully developed operational SCOPE in higher education does not yet exist. It would include operating systems that are not only performing their routines efficiently but are also designed to eliminate duplication and gaps in information gathering and to provide routinely to the MIS the data required for its monitoring function. The latter in turn would produce control reports routinely and supply to the administration the information required for planning and evaluation. The PPB system would have built into it all the essential characteristics previously described and, in addition, a computer-based simulation model to be used as a regular tool in the consideration of the consequences of alternative decisions.

Much progress is being made in the improvement of operating systems, largely through computer applications. Such applications to procedures for student records, accounts payable, facilities inventories, library circulation control, and so on are becoming widespread. They no longer present formidable problems in systems design. Nevertheless, it is also true that in many institutions operating systems have virtually broken down under the pressure of growth; work on management systems has had to be deferred in some cases because of operating emergencies. It is also true that manual systems, both efficient and inefficient, are commonplace, especially in our smaller institutions.

An assessment of the present state of MIS, in universities in the United States is especially difficult because of the confusion of terminology. Many so-called MIS's prove, on investigation, to be simply modernized operating systems dealing with one or two specific sectors, such as facilities inventory or student records. Several universities have in operation a usable but only partial MIS—that is, a system which does provide management data in usable form extracted from several operating systems. The absence of guidelines for systems development, the cost of system design, and programming appear to be the chief constraint at present.

The impetus toward PPB has come largely through government. Several state governments now require that all departments conform to PPB routines, and more have set deadlines for future conformance. Thus the state-supported universities are the ones most heavily involved in making the changes necessary to meet the new requirements. It appears that in some cases the combined effects of time, pressures exerted by the state, and stubbornness on the part of the university have led to conformity in form more than in substance. The chief problems in accommodating education to PPB appear to be the construction of the program budget and the lack of agreement on standards for evaluating the quality of the performance. However, it is accurate to say that most of those who have experienced the PPB discipline agree that the requirements for specifying objectives, for program description, and for explicit consideration of alternatives constitute major benefits to the institution as well as to the funding agency.

With a few notable exceptions, such as the University of California, the work on simulation models has been confined to either a segment of the university (a medical school or a business school) or a specific planning requirement (enrollment or facilities forecasts). However, there is a very widespread interest in large-scale and more general models. If funding is not delayed, as many as eight colleges will have operating simulation models of their own installed in the near future. Additional installations can be expected rapidly thereafter. It may be some time, however, before a general planning model for a complex university, having substantial numbers of sponsored research projects and other activities in addition to instruction, will be operational.

CONCLUSIONS

The following conclusions are based on experiences in working with colleges and universities on their information and planning requirements.

A major impediment to the development of MIS's is the weaknesses in the operating systems. Some of these weaknesses can be corrected fairly easily as a by-product of the development of the MIS. Others are more fundamental and require extensive revision. For example, Student records are not usually organized in such a fashion that accurate measures of attrition can be extracted; yet for purposes of planning and evaluation, student attrition is a matter of major importance. Also, in academic libraries, collections are recorded in numbers of volumes almost everywhere, yet the standards adopted by the librarians themselves refer to title count, which is surely a better indicator of the strength of the collection. And most institutions have developed no measures of library service to users; certainly circulation totals are poor indicators, especially in reference and research collections.

Nevertheless, the solution is not uniformly to work from the "bottom" up, rectifying all the operating systems before moving on to management information, program budgeting, model building, etc., for three reasons. First, the development of operating systems that will later support the MIS requires a good knowledge of what the latter's requirements will be. Second, a serviceable MIS can be developed so long as a sufficient number—not necessarily all—of the main operating systems are in good order. Third, sometimes the need for a serviceable MIS to assist senior administrators is more urgent—to regain control of the operation of the university as a whole, for instance—than the improvement of a single floundering operating system.

Thus the decision as to where to begin is a pragmatic one. The "ideal" plan is not always the best plan; the best plan has the peculiar virtue of being possible. In one case a particular operating system—student records, maintenance of facilities, or classroom scheduling—may be so critically weak that the day-to-day operations of the university are very adversely affected; it becomes the focus of initial effort and the broader aspects are dealt with only when the emergency has been met. In another case the state government may require the implementation of a PPB system by a specified date; the problem then becomes not whether to do it but how to do the best job of it under the circumstances.

It is usually necessary to conduct a review of organization structure of senior administration before devising an MIS. The design of the information system must specify what information must go to whom. If the organization itself is defective, the information system will not correct it. Some responsibilities may overlap or may be ambiguously assigned; these must be clarified. In other cases a function is not assigned at all; in fact, the planning function itself is frequently in such a condition. When the organization structure is not studied and responsibilities not properly aligned, some tasks will be done badly and others not at all. This is a reminder, if it is needed, that an information system is not a panacea.

MIS and PPB systems act as powerful heuristic and recollective devices in their impact upon administrators. The program planning discipline requires that the university leadership confront with great concreteness questions of objectives, evaluation criteria, and priorities which are otherwise easily let slide in the press of daily affairs. Most of our colleges are not yet facing squarely the implications of rising instructional costs and course proliferation, the long-term budgetary impact of debt service on new facilities, the shifts in sources of income. The MIS can and should provide reports which serve as reminders of issues yet to be faced. The student profile may reveal that only economically privileged students are in attendance; the faculty profile may reveal that two-thirds of newly hired teachers are graduates of local institutions. The information itself is "neutral," but the administrator is reminded that he should consider its import.

The capabilities and motivations of administrators are, of course, critical factors. In the United States today the demand for qualified leadership in education far outruns the supply. Among those who have the necessary capabilities there are understandably a significant proportion who are not yet convinced of the value of information systems and the PPB approach. There are at least two ways of overcoming these reservations: deep involvement in the tasks of program definition, and formally organized training programs in which objectives and methods are thoroughly explored.

It is useful to bear in mind that in seeking to identify critical indicators, the search should not be limited to matters in which a judgment is readily possible—such as expenditures in excess of budget or a space utilization rate below standard. The indicators should include data of clearly recognizable importance with respect to which it may be necessary to suspend judgment. Concerning the composition of the student body, for instance, it appears that the United States is about to pass through a second wave of mass democratization of its universities and colleges, perhaps slightly in advance of the first wave in most other countries. (The OECD report, Development of Higher Education in OECD Member Countries: Quantitative Trends [April, 1969], indicates that higher education enrollment in the United States at age eighteen is 34.6 percent of the age group, compared with a range of 0.1 percent to 8.8 percent in other reporting countries; at age nineteen, 33.2 percent, compared with a range of 1.4 percent to 12.5 percent in other reporting countries; at age twenty, 29.2 percent, compared with a range of 3.8 percent to 11.6 percent; and so on.) The university administrator may not be certain what policies he should adopt in the face of this new wave; but it is critically important that he should know how the composition of his student body is changing—the family income, race, and scholastic aptitude of his new students. He will soon have to formulate a policy if he has not already done so, and he had better have the facts.

The single most difficult conceptual problem colleges and universities seem to face in working with PPB is the definition of a "program" in the strict sense of "a group of interrelated activities organized to accomplish an objective." This difficulty stems from the elusiveness of the objectives themselves. Is the objective of undergraduate instruction to produce a specified number of qualified degree holders annually, with so many in each of a number of specified fields of study? In that case, the programs are those courses and other activities necessary to meet the degree requirements. If this line of reasoning is followed, then what is done with the effort expended on students who do not graduate? In the face of these and other difficulties there is a temptation to treat an organizational unit itself as a program; this simplifies the cost allocation but tends to undermine the very foundations of the PPB approach.

Planning reports produced by the MIS must contain substantial amounts of information derived from sources outside the university itself; thus the environmental data must be keyed to the system as fully as the internal operating systems. Six problems contribute to the maladjustment of the universities: financial problems, employment problems (of graduates), curriculum maladjustments (to new disciplines and job requirements), problems of quality, social problems (generated by restrictive admissions policies), and human resource wastages. In at least five of these six cases the measures appropriate to the evaluation of the success of the universities must be drawn in large part from data derived from the economy at large and the society beyond the campus. It is crucial that in the construction of the MIS, standard sources of data be identified with respect to manpower, demography, government expenditures, etc., which can be fed into the information system regularly to provide the subsequent output reports containing appropriate comparisions and ratios relating the university and its graduates to the society at large.

It should be remembered that the object of planning is to allocate scarce resources among competing activities. One of the most painful lessons the United States has had to learn as it has come of age in the world is the limits of its power. Just as the task of foreign policy must be the rationing of one's power among one's aims, so is the task of the university the rationing of its resources among its objectives and programs. The discipline of the PPB system provides constant reminders of these elementary truths. For example, a very small across-the-board faculty salary increase may in fact be a decision to part company with the best faculty members in the university and to hold on tightly to the most incompetent. If constructive rapid change is what society demands of our universities, then it is the task of planning systems to present starkly the hard choices we face.

This suggests, finally, that there is perhaps an even more important objective of planning than the best allocation of resources. That is, the planning process ought to be so organized that it succeeds in raising in virtually inescapable ways the most fundamental questions that must be faced by those responsible for the life of the institution. A good test of the value of the planning process is the amount of discomfort it causes. It should raise all those thorny issues alluded to at the beginning of this chapter: Who is entitled to a place in the university? Is the present faculty suited to the needs of the new students? Are our campuses getting too big? What is the appropriate role of students and of faculty in the governing structure? And so on. How, indeed, can the university decide on the best allocation of its resources until it can answer such questions? And how else can the university transcend the abysmal level of "confrontation politics" to which its enemies would reduce it?

CHAPTER

5

THE WICHE PLANNING AND MANAGEMENT SYSTEMS PROGRAM: ITS NATURE, SCOPE, AND LIMITATIONS

Ben Lawrence

One of the most important current projects aimed at improving the management of higher education institutions is the Western Interstate Commission for Higher Education (WICHE) Planning and Management Systems program, commonly referred to as the WICHE-PMS program. In this chapter, Ben Lawrence, director of the program, describes the operational aspects of WICHE-PMS and the advantages which can be anticipated as a result of this program. Ed.

As the number of persons involved in the Western Interstate Commission for Higher Education Planning and Management Systems program (WICHE-PMS) and the number of projects under way have increased, it has become increasingly difficult for all persons involved to understand the nature, scope, and limitations of the program. It is the purpose of this chapter to explicate them. To do this it is necessary to summarize the goals of WICHE-PMS; identify the basic guidelines presently governing its developments; describe briefly the major functional areas for research and development essential to the improvement of higher education management; identify the research and development areas with which WICHE-PMS is concerned and associate each of its projects (current and planned) with one of these areas; and describe briefly the higher education management process and indicate the point at which each of the projects relates to it.

GOALS

The rapid growth in size and complexity and the current demands for better accountability in higher education have highlighted the need for effective application of new management techniques in colleges and universities. In response to this need, a large number of institutions and agencies of higher education, with the assistance of WICHE, are seeking to develop new management concepts and techniques designed to assist in the improvement of higher education institutional management, of statewide coordination, and of decision-making processes at the highest national levels. (For a more detailed description of the goals, see Appendix A.)

GUIDELINES FOR PROGRAM DEVELOPMENT

There currently exists a set of guidelines for the WICHE-PMS program which have been drawn up with the participation and concurrence of the program's advisory structure. (See Appendix B.) The guidelines may be modified by the advisory structure (particularly the Technical Council) as deemed necessary. Activities that are to be undertaken within the program are determined by the participating institutions and agencies through the advisory structure. (Appendix C describes in detail the organization and makeup of the advisory structure.)

AREAS FOR RESEARCH AND DEVELOPMENT IN THE MANAGEMENT PROCESS

The management process in higher education is complex and interrelated. It occurs at many levels—department, campus, institution, system, state, and nation. It has many commonalities at each of these levels and at the same time has many aspects that are unique to each level. In order to describe the areas in higher education management requiring research and development, a very simplified process model for higher education management is proposed in Chart 4. This model should not be construed as a definitive statement of the process; it is merely one way to describe the process of higher education management. The various elements of the higher education management process can be grouped into four somewhat arbitrary areas that nevertheless represent logical concentrations of research and development activities: goal setting, program planning and resource allocation, execution, and evaluation. There is another area requiring research and development that impinges upon the entire management process and is a necessary foundation to research and development in each of the areas identified above. This additional foundation area may be referred to as the communication base. Figure 1 depicts the five major areas of research and development deemed necessary to the improvement of higher education.

Of course, there is no clear-cut delineation between these somewhat arbitrary areas for research and development related to

CHART 4

A Simple Process Model for Higher Education Management

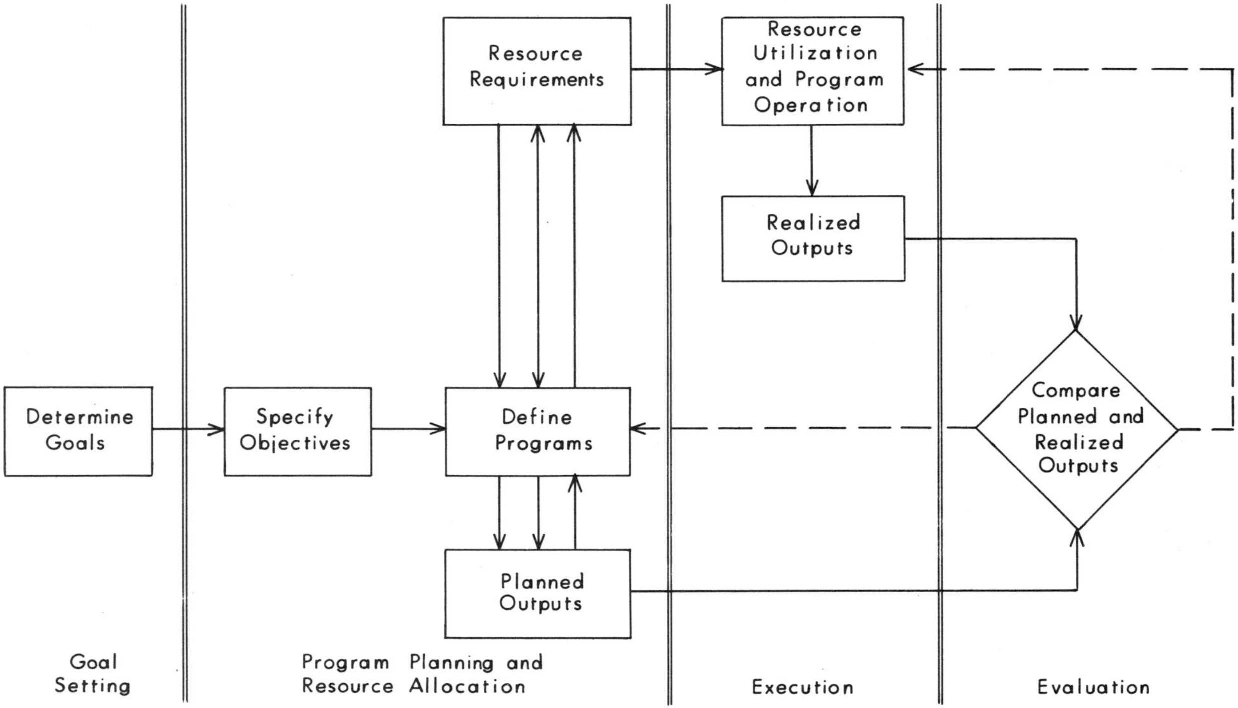

FIGURE 1

Five Major Functional Areas of Research and Development Essential to the Improvement of Higher Education Management

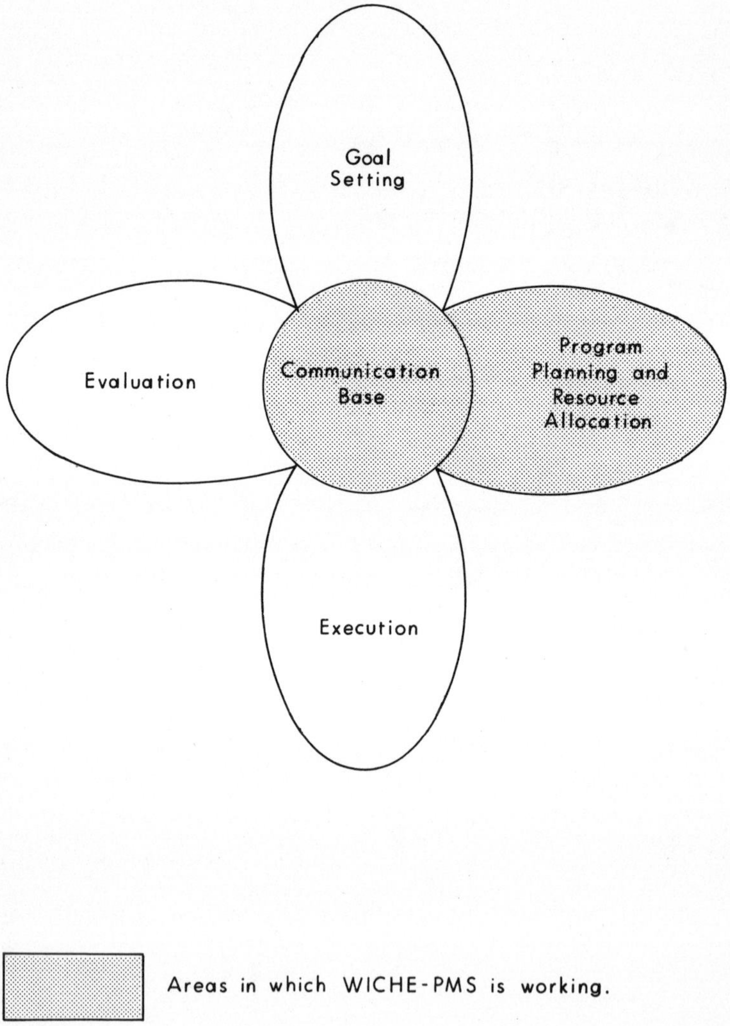

the management process, but describing them in the pure sense helps one to understand better what WICHE-PMS is attempting to do. These areas of research are closely related. The communication base is of central importance because of its important linking function to each of the other major areas.

Goal Setting

Goal setting refers to that activity in which top-level management and the appropriate constituencies—for example, at the state level, the commission for higher education, the legislature, and the public—determine the broad value-oriented purposes of higher education. Goal setting determines "what to do" and "why" in the broad sense. For example, the following statement illustrates the character of a goal.

> The Medical Center of the State University shall conduct research in the health care field and train health care manpower as required to maintain the standard of the state's health care service at the highest level available in the United States.

While established goals are the foundation of the planning process, there is an almost total lack of understanding concerning the methods necessary to the formulation of such goals. Examples of research and development needed in the goal setting area include the following:

1. Studies to consider problems and suggest models of organizational structures for higher education at all levels (department through national) that will facilitate goal setting

2. Research into the processes of determining goals and establishing consensus on these goals

3. Development of exemplary sets of goals for different types of institutions and agencies of different levels that will serve the management function by enabling operational objectives to be developed from those goals.

Program Planning and Resource Allocation

Program planning and resource allocation refers to that function of central or high-level management which is principally oriented toward achieving what we call "economy in the large." This function accomplishes the following:

1. It determines and selects operational (quantifiable) objectives (chosen from among many feasible alternative objectives) necessary to accomplish the goals determined under the goal-setting function.

2. It identifies feasible alternative programs available to accomplish the objectives selected and specifies a set of outputs for each program.

3. It calculates the resources—space, people, equipment, and dollars—necessary to support the respective alternative programs.

4. It evaluates the alternative programs by comparing the planning outputs with expected costs and then proceeds through an iterative process of evaluations and trade-offs until a feasible plan at a reasonable cost is determined. This may ultimately include a reexamination of the overall goals, if the resources do not appear to be available to accomplish the goals as initially established.

Examples of research and development needed in the area of program planning and resources allocation are the following:

1. Development of tools, techniques, and procedures to facilitate the determination of resources required to support specified programs and the examination of trade-offs between alternative resources mixes, between alternative programs, between programs and objectives, between alternative objectives, and between objectives and goals—this generally includes development of methods of generalizing alternative courses of action to accomplish objectives

2. Development of specifications for the measurement of the outputs or benefits of higher education in such a manner as to facilitate the decision-making process

3. Examination of alternative technologies to increase program productivity without loss of quality, e.g., educational television, computer-aided instruction, class size, and curricular mix

4. Development of means to quantify goals in a manner useful to programming and resource allocation.

Execution

Execution is that function of middle management which is concerned principally with achieving "economy in the small." It ensures that what has been chosen to be done is done well and efficiently. It takes the resources allocated and implements the planned program with the intention of producing the outputs as planned—e.g., that secretaries work the required number of hours each day, that supplies are not wasted or stolen, that property is protected, that professors and students get to the right room at the right time, etc.

Examples of research and development needed in the area of execution are the following:

1. Faculty evaluation procedures

2. Personnel evaluation procedures unique to higher educational management processes

3. Design of operational systems related to students, admissions, registration, payroll, etc.

4. Development and evaluation of alternative teaching methodologies or learning situations

5. Organizational unit studies

6. Procedural efficiency studies.

Evaluation

Evaluation is that function of management which is concerned principally with determining the extent to which the organization accomplished what it set out to do. It measures the real outputs or benefits produced by the approved and implemented program and compares them with the planned outputs or benefits. Evaluation requires the development and validation of a host of instruments designed to measure the various outputs of higher education, including the new ones currently being identified and defined.

Communication Base

A major element in the management of higher education at all levels is the use of comparable data. Data must be comparable if it is to be aggregated, exchanged, or incorporated into other equivalent data for reporting and analysis purposes. Comparability is dependent on three factors, regardless of the level or management function: standardized definitions of data elements; standardized procedures for aggregating, reporting, and analyzing those data elements; and agreement on qualitative aspects involved in comparison. Examples of research and development needed in the area of the communication base are data element dictionaries and uniform procedures for aggregating, reporting, and analyzing data for comparative purposes at all levels and of all types.

AREAS OF RESEARCH AND DEVELOPMENT IN WHICH WICHE IS CURRENTLY WORKING

The resources available to WICHE for research and development efforts would not be wisely spent if they were spread across the full spectrum of management. In cooperation with the participating institutions and agencies, those areas considered to be crucial to improved management in higher education were identified. The areas identified for the initial effort of the program are the communication base and program planning and resource allocation.

Table 2 lists the current and planned projects of WICHE-PMS which falls into each of these two areas of research and development. No project can be said to fall exclusively in one category or the other but may be considered to have its primary purpose in that area.

In addition, there are three projects that are designed to apply WICHE-PMS-developed tools at specific levels of management in higher education and to develop tools identified as being particularly needed at those levels. These projects are Small College Demonstration Project (SCDP), Statewide Planning and Management Systems (SWPMS), and Departmental Management Systems (DMS).

RELATIONSHIP OF WICHE-PMS PROJECTS TO THE HIGHER EDUCATION MANAGEMENT PROCESS

Each of the projects in WICHE-PMS (those contemplated as well as those under development) plays a definite role in the development of management systems which are intended for use in higher education at the institutional, state, and national levels. Furthermore, each of the projects is relevant to program planning and resource allocation and requires the kinds of skills being emphasized by WICHE-PMS.

The projects can be grouped into two categories for discussion purposes: projects concerned with the development of common data elements and common procedures (contributions to the communication base previously described) and projects concerned with the development of analytical tools and procedures designed to improve the program planning and resource allocation progress. The categories and the specific projects included within each are discussed below.

TABLE 2

Current and Planned WICHE-PMS Projects

Communication Base	Program Planning and Resource Allocation
1. Data Element Dictionaries (DED)	1. Resource Requirements Prediction Model (RRPM)
2. Program Classification Structure (PCS)	2. Student Flow Model (SFM)
3. Personnel Classification Manual (PCM)	3. Space Analysis Manuals (SAM)
4. Faculty Activity Analysis (FAA)	
5. Input-Output Indicators (I/O)	
6. Cost-Finding Principles (CFP)	
Planned	
1. Information Systems Design Manuals (ISDM)	1. Facilities Costs and Capital Financing (FCF)
2. Information Exchange Procedures (IEP)	2. Higher Education National Indicators Survey (HENIS)
3. Higher Education Finance Manuals (HEFM)	
4. Higher Education Facilities Manuals (HEFA)	

A simple process model for higher education management was depicted in Chart 4; the major areas of the process are described in Chart 5. This chart refers to the same process model but includes an indication of the principal point of impact of each WICHE-PMS project—current and planned—whenever appropriate.

CHART 5

WICHE-PMS Projects Within a Simple Process Model for Higher Education Management

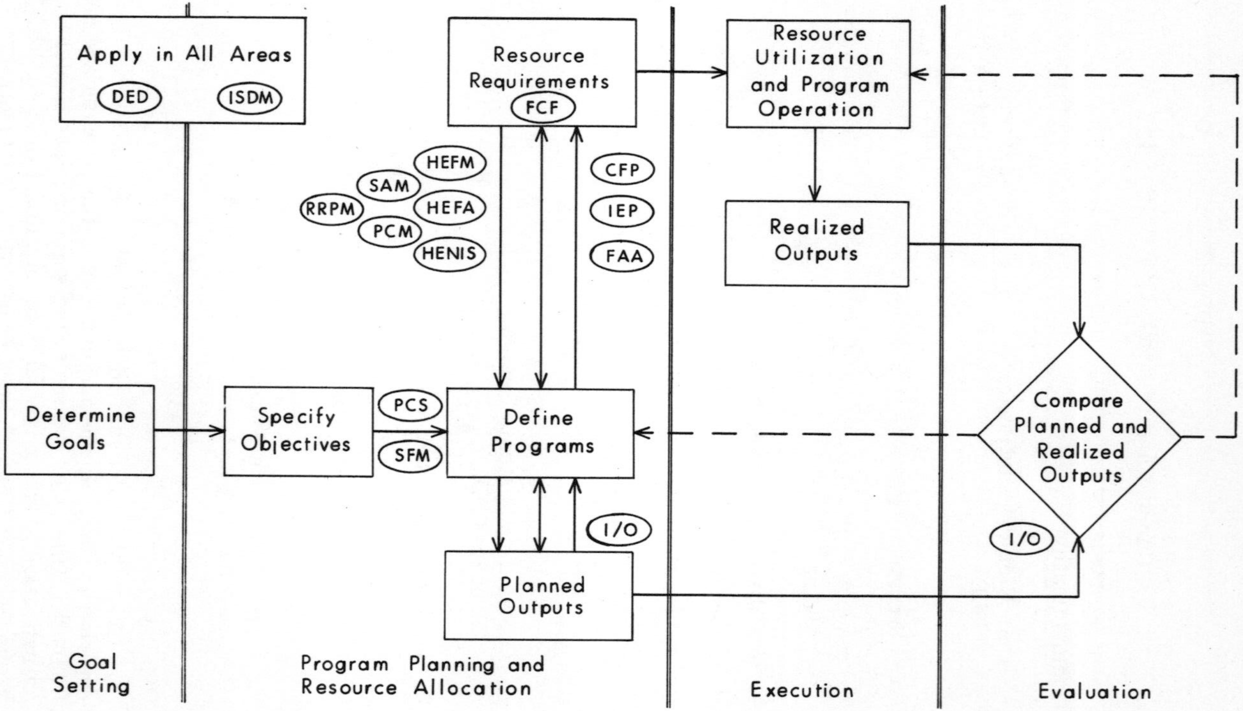

Projects Concerned with the Development of Common Data Elements and Common Procedures

A major element in the management of higher education at all levels is the use of comparable data (especially cost data). At the institutional level, interinstitutional comparisons are currently a particularly important aspect of an individual institution's evaluation process. Since the quality of institutional outputs cannot yet be measured quantitatively, there is no way for an institution to perform cost-benefit analysis on a strictly intrainstitutional basis. In order for institutions to perform such analysis in the absence of quantitative output data, it is common practice for them to identify peer institutions which are perceived to generate outputs of the same quality and to compare their specific cost data with similar data obtained from the peer institutions.

At the state and national levels, comparability of institutional data is even more critical. Since programs at these levels are implemented within institutions, evaluation of these programs cannot begin until data from various sources—different institutions—can be collected and aggregated. In order to perform such aggregation, the data collected must be comparable.

As the need for, and trend toward, increased planning for higher education at the state and national levels continues, the need for a common data base and standard reporting procedures becomes even more important because evaluation of the state and national programs is completely dependent upon the availability of comparable data and information.

Comparability is dependent on three factors: standardized definitions of data elements, standardized procedures for aggregating and analyzing the defined data elements, and agreement on qualitative aspects involved in comparison.

There are ten WICHE-PMS projects concerned with the development of standardized terminology, uniformly defined data bases, and common procedures and conventions.

Data Elements Dictionaries (DED)

DED is designed to define those data elements associated with students, staff, courses, finances, and facilities which are necessary for generating comparable information. For pragmatic reasons, the initial emphasis has been placed on the data interrelationships between the execution and program planning and resource allocation aspects of the management process. DED relates to the entire process of higher education management. It is absolutely essential to the development of comparable information at all levels.

Program Classification Structure (PCS)

PCS has as its objective the development of a common taxonomy of the programs and activities in higher education. The structure will define an exhaustive and mutually exclusive list of higher education's programs as viewed from the institutional level. Since programs are the common factors which run through all elements of the management function, the PCS is of primary importance. It establishes the framework for all of the other projects within WICHE-PMS. PCS relates to the management process in higher education at the institutional level, primarily at the point where the program is defined (indicated by PCS on the model in Chart 5). However, it affects the entire process, for it is a necessary basis for the development of comparable information.

Personnel Classification Manual (PCM)

PCM is concerned with developing a classification system for higher education personnel and with preparing a glossary of personnel terms. It is designed to expand the state of the art and to develop a system of personnel classification which fits within the framework of PCS. PCM provides the bridge by which data concerning personnel (as defined in DED) are associated with programs (as defined by PCS). It relates to the higher education management process at the point where resources are associated with programs, as indicated by PCM on Chart 5.

Cost-Finding Principles (CFP)

CFP is concerned with developing conventional procedures for allocating all costs to specified programs within an institution. In particular, this project is designed to describe the procedures for allocating costs in such a way that the total costs of operating each of the programs can be determined. The determination of cost-finding principles is a necessary step in the eventual development of comparable cost information. This project also relates to the higher education management process at the point where resources are associated with programs, as indicated by CFP on Chart 5.

Faculty Activity Analysis (FAA)

FAA is designed to develop standardized procedures for analyzing faculty activities and to establish conventions for allocating faculty resources to the institutional programs. It is concerned primarily with developing common procedures to be used in sorting out the faculty resources devoted to various programs. As such, FAA is a fundamental component in the development of comparable cost data. Since personnel costs are the largest single item in most programs, it is imperative that those procedures developed to determine the cost of each program accurately reflect the allocation of personnel resources to programs. This project, therefore, is intended to aid in the further development of CFP and the Information Exchange Procedures project. FAA relates to the higher education management process at the point where resources are associated with program, as indicated by FAA on Chart 5.

Higher Education Finance Manuals (HEFM)

HEFM is designed to develop procedures for aggregating institutional financial data, as now maintained, into the formats required by the use of program planning and management systems at the institutional, state, and federal levels. In the main, HEFM is designed to help the institutional administrator rearrange the financial data currently being collected into formats more specifically oriented to the needs arising from the use of planning and management systems methodologies at all levels of higher education. HEFM relates to the higher education management process at the point where resources are associated with programs, as indicated by HEFM on Chart 5. This project is relevant at all levels of higher education management.

Information Exchange Procedures (IEP)

IEP is designed to develop and test procedures for producing comparable data from institutions of higher education. The data to be produced will include, but will not be limited to, cost of instruction, by level of student and field of study; cost of instruction, by level of course and discipline; cost per unit of output, e.g., certificate or degree.

This project is, in essence, the mechanism by which outputs of several other projects (PCS, HEFM, CFP, and FAA) are orchestrated and interrelated. Given the specific procedures developed within these other projects, the procedures specified by IEP will describe conventions concerning the ways in which the pieces will be woven into the information to be exchanged between institutions. This project also affects the area where resources are associated with programs, as indicated by IEP on Chart 5.

Input-Output Indicators (I/O)

As currently envisioned, I/O is intended to develop a taxonomy of the outputs and activities of higher education, to suggest possible means of obtaining quantitative measures of these outputs, and to encourage the development of such measures. Inasmuch as WICHE-PMS is not stressing expertise in the behavioral sciences, the development and validation of tests which measure these outputs are not considered as being within its area of greatest competence. By virtue of the nature of the program planning and resource allocation process, however, it will be necessary for WICHE-PMS to be actively involved in determining which outputs should be measured in order to provide the necessary information for decision making. This project relates to the management process in higher education at two points—where planned outputs are postulated and where realized outputs are evaluated, as indicated by I/O on Chart 5.

Higher Education Facilities Manual (HEFA)

The objective of HEFA is to produce a revision of the Higher Education Facilities Classification and Inventory Procedures Manual. The first edition of this manual was published by the U.S. Office of Education in 1968. It has since been used as the basis for the reporting of facilities data through the Higher Education General Information Survey. Widespread use of the material included in this document has pointed out several areas requiring revision and more explicit definition. In addition, a major element in the revision will be fitting the facilities elements within the framework of PCS. The manual provides uniform procedures for classifying and inventorying space within the institution. This project also affects the management process at the point where resources are associated with programs, as indicated by HEFA on Chart 5. This project is relevant at all levels of higher education management.

Information Systems Design Manuals (ISDM)

The objective of ISDM is to develop and publish a comprehensive manual dealing with the concepts underlying the design, installation, and operation of information systems for institutions of higher education. Its four sections will cover concepts of information management, concepts of data systems, concepts of installation management, and higher education applications. ISDM relates to the entire process of higher education management and is imperative to the provision of readily accessible and easily analyzed data from the operating information systems to the program planning and resource allocation portion of the management process.

Projects Concerned with the Development of Analytical Tools and Procedures

In addition to those projects which deal specifically with the standardization of data elements and procedures, WICHE-PMS is engaged in several projects which have as their objective the development of certain tools which can be used to improve program planning and resource allocation. Whereas the projects in the previous category were specifically oriented toward the present or the immediate past, the projects concerned with the development of models and other tools are oriented toward the future. The models are intended to aid in the investigation of the relationships between variables and possible future effects as one or more of these variables are changed. The role of these projects, with respect to the entire subject of planning and management systems for higher education, can best be explained by reference to Chart 5.

Student Flow Model (SFM)

SFM is designed to develop analytical models which will aid in predicting student enrollments and will simulate student progression through postsecondary education. At the institutional level, such a tool is necessary to an investigation of how a student body will become distributed over the various programs being offered, and of how the institution can act in an effort to obtain the numbers and types of students desired. An institutional SFM will aid in predicting the levels of activity within each of the institutional programs—SFM on Chart 5. A state-level SFM will aid in predicting the distribution of students over institutions within the state.

Space Analysis Manuals (SAM)

The objective of SAM is to write a series of manuals which describe and illustrate methods and procedures for use in the planning and management of college and university facilities. As part of the planning process, this project is designed to provide methods of estimating the various types of facility resources required to support the different programs at the projected level of activity. As part of the management process, they are designed to aid in the allocation of existing facility resources to the current institutional programs. This project is intended to bridge the gap between programs and space resources, as indicated by SAM on Chart 5.

Facilities Costs and Capital Financing (FCF)

The objective of FCF (a proposed project) is to investigate the factors influencing construction and plant operating costs in colleges and universities. As an outcome of these investigations, methods will be developed for estimating construction and plant operation costs, analyzing the trade-offs between additional capital expenditures and operating costs, and allocating capital and plant operating costs to institutional programs for use in the analytical models and information exchange procedures being developed. As indicated by FCF on Chart 5, this project is designed to bridge the gap between space resources expressed in square feet and space resources expressed in dollar terms. It is also intended to aid in investigating the educational production function.

Resource Requirements Prediction Model (RRPM)

RRPM is designed to develop an analytical model of the relationships between programs and levels of activity and resources required within programs. Inasmuch as planning is an iterative process, a tool which allows an administrator to determine quickly the resource requirements of different courses of action is extremely important. RRPM is intended to be such a tool, being the mathematical expression of the relationship between programs (and levels of activity within programs) and associated resource needs. RRPM relates to the management process as indicated by RRPM on Chart 5.

Higher Education National Indicators Survey (HENIS)

HENIS is intended to develop such things as lead-lag indicators for income and expenditures for higher education (i.e., indicators which will function much as the consumer price index functions for the economy as a whole). The tools developed by such a project will be designed to aid decision makers at the national level in testing the feasibility of pursuing certain programs (i.e., will it be economically feasible to undertake certain programs?). This is the only project currently being contemplated which will not deal primarily with specific elements of an institution's decision-making process. It will draw upon the information and information exchange procedures developed by WICHE-PMS.

Additional Projects

Besides those projects which can be readily grouped into one of the two previous categories, WICHE-PMS has proposed three projects which are designed for use by special audiences within the higher education community.

Small College Demonstration Project (SCDP)

SCDP is intended to identify the products of WICHE-PMS which could be most appropriately adapted for application in small colleges; make whatever modifications in these tools are necessary to meet the special needs of small colleges; implement these tools and procedures in three to five small colleges; and provide mechanisms for displaying the results to representatives of other small colleges.

Statewide Planning and Management Systems (SWPMS)

SWPMS covers a wide scope of activities, from development of tools to aid in establishing goals and objectives for higher education at the state level to development of procedures to aid in investigating the sensitivity of institutional programming to programs at the state and national levels. Particular attention will be devoted to the integration of state and federal processes with those which occur within the institution. This project relates to the higher education management process at the point where resources are associated with programs and objectives.

Departmental Management Systems (DMS)

DMS will address those elements of the higher education management process which are of particular concern to department chairmen. Tools and procedures will be developed which will help department chairmen in areas such as allocation of faculty resources to particular activities and investigation of the implications of alternative instructional technologies (lecture, closed-circuit television, computer-assisted instruction).

APPENDIX A: GOALS OF WICHE-PMS

WICHE-PMS is a cooperative project among higher education institutions and coordinating agencies to design, develop, and implement management systems and data bases, including common uniform data elements.

The rapid growth in size and complexity of higher education has highlighted the need for systematic collection and use of data in the effective management of colleges, universities, and state systems of higher education. Without systematic, accurate feedback to management of the effects of its operations, an institution or system can waste its resources on ineffective or unnecessarily costly activities. However, judgments about effectiveness and relative costs cannot be made adequately in isolation. Therefore comparable data from other organizations of similar complexity and with similar missions are needed.

In meeting these common needs, state coordinating agencies and concerned colleges and universities have asked WICHE to bring together a highly competent staff to develop new planning and management concepts and techniques designed to assist in the improvement of higher education institutional management, of statewide coordination, and of decision-making processes at the highest national levels.

The achievement of these goals will be accomplished through the design, development, and implementation of planning and management systems, which will be based on information systems that have common data elements. These systems will be designed to benefit the entire range of higher education organizations, including public and private community colleges, four-year institutions, and universities, as well as local, state, regional, and national agencies.

The purpose of the planning and management systems and data bases in relation to the achievement of the three stated goals will be the following:

1. To promote the development of various aids to more knowledgeable resource allocation within institutions and agencies of higher education

2. To enhance the ability of institutions and agencies to exchange and report comparable information

3. To assist the participating institutions and agencies to implement the products of this program through a variety of training seminars and materials.

At the request of the cooperating organizations, the initial phases of this project are concerned with establishing a preliminary set of common data elements, program elements, program classification structure, program budget categories, and quantitative indicators: designing and implementing compatible management systems which will provide uniform data from which the costs of instructional programs by level of course, level of student, and field of study can be derived. To be eligible for inclusion in the project, institutions must be willing and able to cooperate in these activities.

Subsequent phases of the project will be devoted to the following tasks:

1. Refining the planning and management systems and data elements, input-output indicators, and cost relationships of the instructional program

2. Beginning the task of identifying cost and benefit indicators of research and public service programs

3. Disseminating information about the project and developing procedures whereby institutions and agencies participating in the project can benefit

4. Performing maintenance operations on the systems which have been developed so as to reflect subsequent progress and developments.

The following missions are essential to the successful achievement of the project goals and purposes and also will be undertaken by the project staff:

1. Stimulating, coordinating, and conducting educational programs at various levels for all institutions and agencies that wish to develop their capability to cooperate in this project. This will include inter- and intracampus seminars in systems analysis, operations research, program budgeting, and cost-benefit analysis; the use of simulation models for training high-level management in the application of these decision-making tools under a variety of institutional circumstances; and the publication and distribution of staff technical reports developed in the process of establishing data definitions, program elements, system applications, input-output indicators, and program budget categories

2. Coordinating development of data elements and information systems at regional and national levels, through cooperation with other regional and state level organizations

3. Coordinating the exchange of comparable higher education management data among the cooperating agencies and institutions in the nation.

The initial mission of the project is to establish the mechanisms by which these objectives will be achieved. A central professional staff will be headed by a director with extensive experience at a high level in the management of higher education. He will be assisted by a staff with substantial experience and demonstrated ability in developing planning and management systems and the application of computers to their operation. The staff will be assisted by appropriate specialists employed by WICHE or contributed by the cooperating organizations as they are needed. An advisory structure, representative of the organizations cooperating in the project, will monitor and guide the work of the central staff.

APPENDIX B: GUIDELINES FOR PROGRAM DEVELOPMENT

Initial Emphasis: Cost of Instructional Program

The initial phases of the program will be concerned with the development of tools to aid in deriving information about the costs of student instruction, including those aspects of research and external service which are directly related to costing the various elements of student instruction. From this initial emphasis the project will move forward to include all categories of information relevant to resource allocation on the campus and state systems.

Universality

The most important guideline is that the set of data which is collected must be of the greatest practical universality and flexibility, so that all levels of institutions and any individual institution can use, on a common and consistent basis, those parts of interest to them beyond the requirements for participation. Similarly, allowance must be made for suitable aggregation of the data so that they may be used for review purposes at echelons above the campus level.

Expandability

The program started with a small nucleus of institutions and agencies which have information systems and are ready to adopt the common data elements required for deriving instructional cost data. At the same time, a larger number of institutions and agencies participated with the staff to develop management systems with common data elements. During the past year, 1970, these institutions were added to the nucleus. During the current phases of the program deliberate efforts will be made to expand the network to include any institution or agency wishing to participate, using the various products developed by the program.

Comparable Data

The data in the system should be capable of being compiled in a variety of formats for reasonably comparable results without requiring all institutions to go to a single operating system. This requires common data element definitions (or translation algorithms) so that institutional information can be incorporated into various studies for comparisons of costs and effectiveness. Wherever possible, existing definitions should be used.

Framework for Exchange

It has proved necessary to develop a standard classification of institutional activities which will provide the framework for information exchange. This classification system is referred to as the WICHE PCS. It provides a consistent means of identifying and organizing the activities of higher education in a program-oriented manner. Together with the WICHE DED, the PCS provides the basic foundation for establishing a framework to facilitate the exchange of management information among institutions and agencies of higher education.

Exchange of Data

For internal administrative purposes and external policy development, institutions and agencies will inevitably compare programs. A common reporting framework will increase the ease and sophistication of such comparisons. Clarifying the relationships between program activities, costs, and objectives can help avoid the now frequent "apples and oranges" errors in comparative studies.

Frequently unnecessary amounts of work are involved because of the slightly different form of data requests from another institution, a state agency, or the U.S. Office of Education. The implementation of compatible systems of data reporting will alleviate this situation to a significant degree.

Data Requirements for Management Decisions

Requirements for resource allocation decisions entail a need for specialized data systems which rapidly, accurately, and routinely provide information to managers at all levels to assist them in making better decisions about preferred courses of action. The development of analytical models will help to identify and understand the implications of alternative policy decisions.

Institutional Programs and Program Budgeting

Instruction, research, and external service are the three major program categories of universities, while liberal arts colleges focus primarily upon instruction. Supporting program elements are libraries, plant maintenance and operation, student services, general administration, institutional research, organized activities, and auxiliary enterprise activities.

A useful tool for higher education will be the program budget, which relates program activities to institutional goals and highlights the principal resource allocation decisions.

The building blocks of major programs must be uniformly defined in a manner which will allow individual institutions to organize them as they wish, yet allow reorganization and comparison by institutions which organize their programs in a slightly different manner.

Institutional Goals

Colleges and universities have multiple goals. It is necessary for the administration of an institution to articulate institutional goals and to do so in a manner which facilitates the relating of program activities and resources to objectives. This is basic to the program budgeting and cost-benefit approach. These relationships should also be made clear in an academic plan for the institution.

Faculty Work Loads

Reasonably accurate data on salaries paid professors is now collected, but there is incomplete data on faculty work load. New tools are needed to help determine the manner in which faculty resources are utilized to accomplish institutional objectives and the relation of faculty expenditures to program costs.

Cost-Benefit, Cost-Effectiveness Analyses

Management needs to know the relationship between costs and whatever criteria of effectiveness or benefit are established as the basis for efficient allocation of resources. The information systems will provide these data. It should be stressed, however, that the data and their analysis do not establish the effectiveness of benefit criteria.

The difference between inputs and outputs may be interpreted as "value added" or "benefits" to the student, to knowledge, or to the public. Relating the amounts of resources used in a program to the difference between inputs and outputs may possibly yield a cost-benefit index.

At this stage there is little agreement as to exactly what should be measured, or how, when considering the outputs of a program. Value added is a concept important to determining cost-benefit. While it is difficult to identify and measure, the attempt must be made. Is it true, for example, as some research findings suggest, that the quality of an institution's graduates is determined largely by the quality of student it admits, rather than by "adding values" to the person during his student years?

Competent Staff Analysis: The Key

The development of planning and management systems will be of little benefit to an institution or agency unless a trained staff is available to conduct appropriate analyses for administrative decision making. It is recognized that many institutions are not now adequately staffed for these functions. Therefore, it is necessary to establish a comprehensive training effort to accompany the systems developments and to urge strengthening of this capability along with the systems effort.

APPENDIX C: ADVISORY STRUCTURE FOR WICHE-PMS

As WICHE-PMS enlarges in scope, reorganization of the advisory functions is necessary for the successful operation of the program. Formal communication lines between the staff and the program's constituency must be developed and maintained in order to provide a continuing basis for obtaining general, policy, and technical advice from the larger community. An advisory structure has been established to accomplish these aims.

Advisory Council

The advisory council will provide communication between the program staff and the participating institutions and agencies. Useful advice can result only when the participants are adequately informed of the progress of the program and when interaction among all participants is fostered. The advisory council will provide the basis for maintaining liaison between the staff and the various segments of higher education that are participating in the program.

Its responsibilities will be to provide a general forum in which participants can discuss the status and development of management systems in higher education; a sounding board for those who are responsible for the policy planning and program development of management and information systems in higher education; and a mechanism for the timely communication of developments within WICHE-PMS.

Membership of the advisory council will include former WICHE MIS steering committee members (as of its last official meeting before dissolution); a representative (chief liaison officer) from each of the institutions, agencies, or higher education systems registered as participating in WICHE-PMS at levels III and IV; and invited observers, as specified by the director of the PMS program. An institution or agency will be represented on the advisory council as long as the participant remains at level III or IV. The representative (chief liaison officer) is designated by an appropriate officer of the Level III or IV participant.

The advisory council will meet twice annually, the time, place, and agenda of the meetings being determined by the executive committee. The chairman of the executive committee will preside over advisory council sessions. Unless the chairman rules otherwise, decisions will be reached by consensus. The chairman will state each decision and the major points of discussion leading to the decision so as to afford the opportunity for contrary professional opinion to be recorded.

National Advisory Panel

The national advisory panel will provide advice from national and regional organizations concerning national priorities for PMS development; relationships between WICHE-PMS and similar or related efforts elsewhere in the nation; and program objectives, guidelines, and procedures.

Members will be appointed by the executive director of WICHE in consultation with the director of PMS. The term of membership is two years, beginning on July 1. Appointments will be made annually on June 1. The terms are staggered in order that each year approximately half of the membership is rotated. Members may be reappointed. The following categories are represented: higher education associations; national professional organizations that have an interest in, or concern for, higher education planning and management systems; and regional and national compacts for education. The chairman will be appointed annually by the executive director of WICHE in consultation with the director of PMS.

The national advisory panel will meet twice annually; the time, place, and agenda to be determined by the chairman and the director of PMS. Unless the chairman rules otherwise, decisions will be reached by consensus. The chairman will state each decision and the major points of discussion leading to the decision, so as to afford the opportunity for contrary professional opinion to be recorded.

Executive Committee

The executive committee will advise the director concerning the general development and direction of WICHE-PMS. Specifically, it will provide advice concerning major priorities for program development; establishment of coordinated and cooperative efforts for PMS development; relationships between the program, its participants, and other planning and management systems developments; major committee appointments; and selection of pilot institutions.

The executive committee will have at least twenty, and no more than twenty-five, members; they will be appointed by the executive director of WICHE in consultation with the director of PMS. The term of office is two years, beginning on July 1. Appointments will be made annually on June 1. The terms will be staggered, in order that each year approximately half of the membership is rotated. The following (not necessarily mutually exclusive groups) are represented: the PMS advisory council, the PMS national advisory panel, the PMS technical council, each major geographic region or major regional compact for higher education, statewide coordinating agencies, community colleges, multicampus systems (central office), private four-year colleges, private universities, public four-year colleges, and public universities. The executive committee members will be appointed in such a manner as to ensure that approximately half of the membership represents institutions of higher education (i. e., colleges and universities), with an adequate number of representatives from the private sector. The remaining members of the committee are to be representative of the various agencies and associations of higher education, with an adequate number of representatives from statewide coordinating agencies. The chairman of the executive committee will be appointed annually by the director of PMS.

The executive committee will meet approximately four times a year; although additional meetings may be called by the chairman. The time, place, and agenda of the executive committee meetings will be determined by the director of PMS, in consultation with the chairman. Unless the chairman rules otherwise, decisions will be reached by consensus. The chairman will state each decision and the major points of discussion leading to the decision, so as to afford the opportunity for contrary professional opinion to be recorded.

Technical Council

The technical council will provide advice to the staff on matters concerning the general technical development of the program. They will be chosen primarily for their expertise and will provide advice concerning the technical boundaries for planning and management systems; the technical priorities and feasible alternatives for the program; major programmatic recommendations for the technical council being reviewed by the executive committee; general concepts underlying specific analytical tools and studies; the selection of contractors, major staff appointments, special projects, and research undertakings; selection of institutions for pilot testing centers; and major committee appointments.

The technical council will consist of no more than twelve appointed members and four ex officio members. Members who are technically qualified will be appointed by the executive director of WICHE in consultation with the director of PMS. Ex officio members are the director of PMS, the director of the research program, the director of the development and applications program, and the director of the training program. The chairman of the technical council will be appointed by the executive director of WICHE in consultation with the director of PMS.

The technical council will meet six times a year; and additional meetings may be called by the chairman. The time, place, and agenda will be determined by the director of PMS in consultation with the chairman. Unless the chairman rules otherwise, decisions will be reached by consensus. The chairman will state each decision and the major points of discussion leading to the decision, so as to afford the opportunity for contrary professional opinion to be recorded.

Task Forces

Task forces will provide consultation and guidance for each of the projects within the program. Their responsibilities will be determined by each project analyst.

Task forces will have at least five and no more than eight members, to be appointed by the director of PMS. Membership nominations will be solicited from the level III and IV participants, unless special circumstances necessitate other nominations. Membership on a particular task force is temporary and may be rotated to provide a range of expertise as required by the specific project.

Task forces will meet as necessary, at the request of the project analyst.

Amendment

The advisory structure for PMS may be amended at any regular meeting of the WICHE commissioners.

When a member of an advisory group terminates his position with an institution or agency, his membership on the advisory group in question is automatically terminated. The executive director of WICHE and the director of PMS, who are responsible for selecting his replacement, may consider him for reappointment.

CHAPTER

6

PROJECT PRIME: A TEST IMPLEMENTATION OF THE CAMPUS SIMULATION MODEL

David C. Cordes

PRIME is an acronym for Planning Resources in Minnesota Education. Project PRIME was a one-year (ending June 30, 1971) research effort jointly funded by the Minnesota state college system, the Minnesota junior college system, the University of Minnesota, the Hill Family Foundation, and the Minnesota Higher Education Coordinating Commission. The project's primary objective was the test implementation of the CAMPUS simulation model in one state college (Bemidji), in one junior college (Lakewood), and in one school at the University of Minnesota. Two other units were added—the division of educational psychology at the University of Minnesota and Hopkins school district. This chapter records the experiences gained from Project PRIME and points out some of the pros and cons of comprehensive simulation models of educational institutions. Ed.

CAMPUS is an acronym for Comprehensive Analytical Methods for Planning University Systems. It was developed under a Ford Foundation grant by the Institute for Policy Analysis at the University of Toronto, which has an extensive research program called Systems Analysis for Efficient Resource Allocation in Higher Education. The program consists of six integrated projects: (1) program planning and budgeting in universities; (2) planning and financing higher education; (3) models for university planning (CAMPUS); (4) integrated university information systems; (5) models for planning and use of physical facilities; and (6) planning and management systems for university information resource centers.[1]

DEVELOPMENT OF CAMPUS

CAMPUS development began with a pilot study in December, 1964. The model has evolved through five revisions, with CAMPUS V becoming available to the public (at a nominal cost of approximately $100) in early 1970. A historical perspective on the evolution of CAMPUS I-IV was published in 1969.[2]

Another version of CAMPUS, labeled either CAMPUS VI or CAMPUS CONNECT is available from the Systems Research Group.[3] The principal difference between CAMPUS V and CAMPUS VI is that the latter allows the user to interact with the model. The basic approach to resource modeling, data structuring, and simulating are similar in both models. CAMPUS V will be used in all succeeding discussion in this chapter.

The journal articles and speeches available describing CAMPUS are numerous.[4] There is also documentation available with the system.[5] However, these reports and available documentation do not describe CAMPUS in sufficient detail to assess its capabilities and limitations.* Questions such as the following cannot be answered from the available literature: How large an institution can CAMPUS V handle? What capability does the model have for converting budget data to program element data? To what level of detail can CAMPUS "represent reality?" How does CAMPUS construct a degree program? How does it handle the support programs?

Project PRIME

Project PRIME was an initial attempt to answer management-oriented questions like those above. The purpose of this chapter is to describe what occurred during PRIME's year of existence and from this experience try to glean, in retrospect, the value of using a simulation model like CAMPUS in planning higher education.

In order to appreciate the test implementation, three background considerations are relevant. First, a basic comprehension of the

---

*The reports tend to describe what CAMPUS "might" do, or what is "planned." This is not meant as a criticism, for the development of CAMPUS is an ongoing effort. The documentation describes the computer logic in detail but does not answer the "management"-oriented questions.

CAMPUS simulation model is necessary. Second, it is important to appreciate the institutional settings within which the test implementations were attempted. Third, it is necessary to review briefly the resources and secondary objectives of the project.

With the above background, the reader can better appreciate the second section, which analyzes the results at each institution. The third section describes several changes made in the CAMPUS model to develop a ten-year program-oriented budget. The fourth provides an introduction to the reports available from the project. The last section presents conclusions and recommendations pertaining to the use of simulation models for improving planning in higher education.

The CAMPUS Model

CAMPUS V is a computer-based resource simulation model with an ability to "represent reality" in considerable detail. As an indication of this level of detail, each course (or activity) in CAMPUS requires the following resource data:

1. Staff—academic (e.g., professor) and academic support (e.g., teaching assistant)

2. Space—classrooms, instruction labs, or special labs

3. Teaching Equipment—projectors, Bunsen burners, computers, etc.

4. Type of Course—lecture, lab, individual study, etc.

5. Staff Specialization—accounting, philosophy, art, etc.

6. Schedule Time—hours per class and classes per week

7. Success Factor—probability of completing courses with a passing grade.

Before continuing this overview of the CAMPUS model, it is necessary to define key management variables.

Cost Center

The cost center is a unique subset or area of the organization being modeled for which resources are required. The prevalent cost center in higher education is the academic department. Other cost centers include research centers, libraries, and deans' offices. An example of a cost center structure is shown on Chart 6, which has ten cost centers at three "levels" (cost centers 5-10 are all level "1").

Program Structure

Program structure is a grouping of the organization's activities and programs in a manner that indicates their relationship to the organization's goals and objectives. Chart 7 is a sample program structure.

Activity

An activity is an event or action that requires resources from a cost center. The typical activity in higher education is a course or section. Other examples of activities include advising students on dissertations and conducting oral exams.

Student

In CAMPUS a student is one person taking one or more courses. "Head count" would be a synonymous term.

Enrollee Load

One enrollee load is one "student" taking one activity (course). Thus, an enrollee load of five could then be either five students taking one course or one student taking five courses.

Affiliated Students

The term "affiliated students" refers to a head count of students "affiliated" with a particular cost center (e.g., a mechanical engineer in the mechanical engineering department).

Affiliated Enrollees

Affiliated enrollees provide an indication of load put on a cost center by its affiliated students.

Academic Staff (Full-Time)

The full-time academic staff is the professional staff at academic cost centers considered in the hiring and promotion policies. Typical ranks include professor, associate professor, assistant professor, instructor, and lecturer.

CHART 6

Cost Center Structure for a School of Business Administration

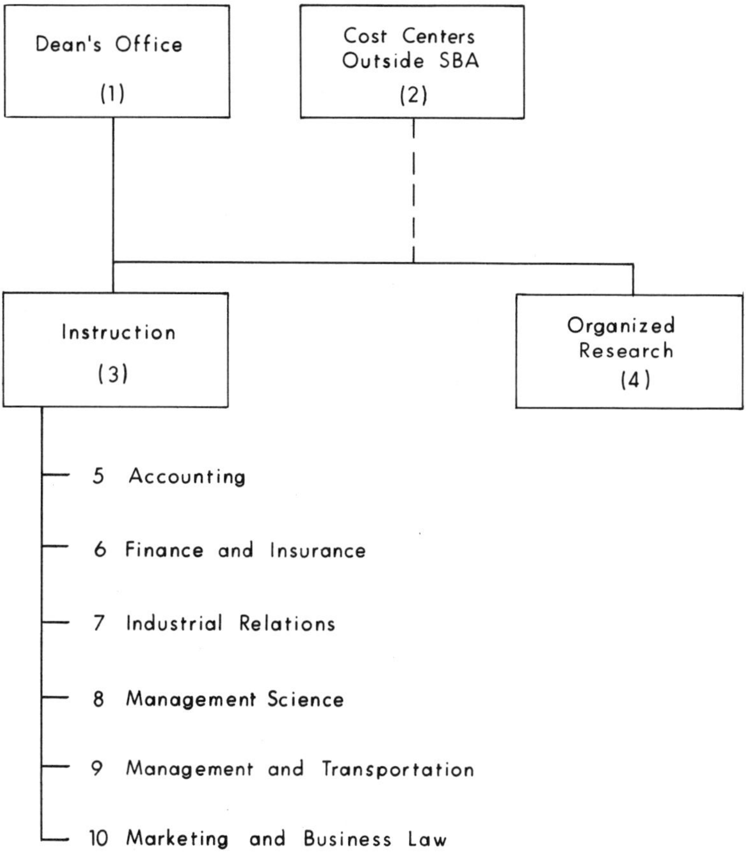

CHART 7

Illustrative Program Structure for a School of Business Administration, Based on WICHE-PMS Format

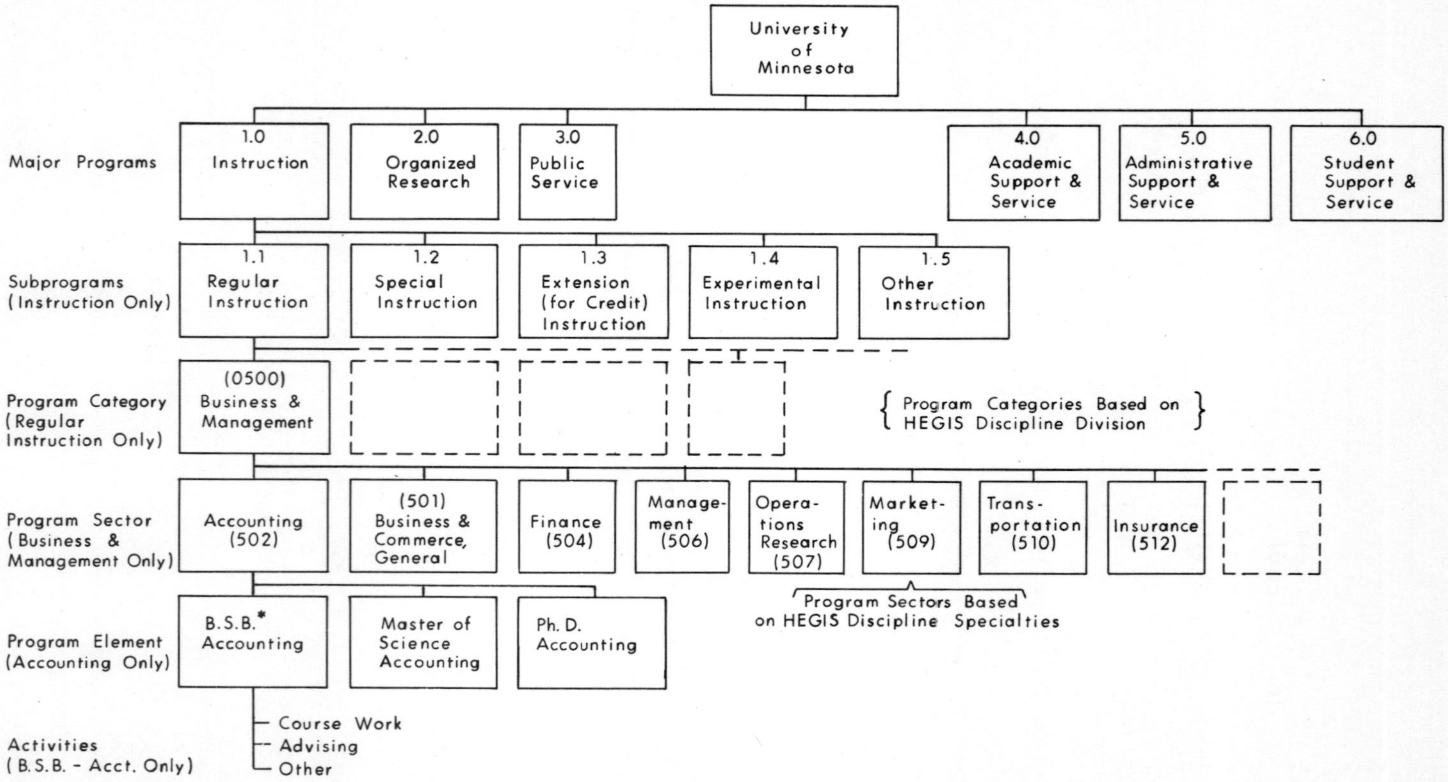

*B.S.B. = Bachelor of Science in Business

Source: Gulko, 1970.

Academic Staff (Part-Time)

Part-time academic staff participates in teaching activities but is not considered in the hiring and promotional policies. A typical rank would be teaching associate.

Academic Support Staff

Academic support staff participates in teaching activities but does not have control of the course. This staff is typified by teaching assistants, graders, and tutors.

Nonacademic Support Staff

Nonacademic support staff is typified by secretaries, research associates, research assistants, and "civil service" staff located in academic cost centers.

Service Department

The service department is a nonacademic subunit within the organization that has requirements and responsibility for resources, e.g., staff, equipment, space. Examples include a computer center, placement office, etc.

Simulation Period

The simulation period may be either a quarter or a semester.

Session

A session is an academic year consisting of a number of simulation periods (i.e., two for a semester system).

Functional or Proportional Basis

The functional or proportional basis is a procedure in CAMPUS for determining the amount of required resources as a function of various measures of cost center activities. An example of a possible function for determining the required amount of computer supplies for the accounting department might be as follows:

computer supplies (accounting dept.) = \$20,000 + \$100 per student + \$500 per faculty member

CAMPUS would then determine the number of students (e.g., 100) and number of faculty members (e.g., 10) for this particular time period (say fall quarter). For example:

computer supplies (accting. dept., fall quarter, 1969/70) = \$20,000 + \$100(100) + \$500(10) = \$35,000

There are fifty indicators of cost center activities available to CAMPUS. These indicators, known as functional or proportional bases, are listed in Table 3.

The Command Structure in CAMPUS

The basic command structure in CAMPUS is displayed schematically in Figure 2. There are three levels of "commands," but only the six level 1 commands are shown on the figure: INPUT, DATA, REPORT, SIMULATE, EXPERIMENT, FINISH. There six level 1 commands control the overall running of the CAMPUS model.

Inputs

The best procedure for understanding the CAMPUS model is to study the inputs in great detail. In this way the reader can develop a "feel" for how CAMPUS operates and at what level of detail. Table 4 outlines the level 2 and level 3 commands associated with the level 1 command INPUT. These sixty-eight level 3 commands (e.g., DEFINE 01, ACTIVITY 01) are the specific card formats required by the model. Each of the fourteen level 2 commands is outlined briefly below.

Define. This level 2 command "defines" or structures the institution's programs and cost centers (departments).

Activity. Activities (courses) are related to programs (degrees) through the use of participation rates. This relationship will be explored in depth below. Also established are the lengths of the degree programs (two years, four years, etc.) and the numbers of credits needed for graduation.

Student. Both freshmen and entering advanced-standing students are input in this section. Dropouts and transfers from major to major are also entered in this section.

Staff and X-Staff. Staffing units required for each teaching and non-teaching duty are entered here, plus salary and office space. Hiring and promotion policy variables are also part of this input.

TABLE 3

Functional Bases for the Calculation of Indirect Resources at a Cost Center

Absolute Functional Bases		
Absolute Value	.01	(28)
Absolute Value	.1	(27)
Absolute Value	1.0	(1)
Absolute Value	10.0	(29)
Absolute Value	100.0	(30)
Absolute Value	1000.0	(31)

Student-Related Functional Bases	
Affiliated Students	(2)
Affiliated Students (in hundreds)	(42)
Aggregate Affiliated Students	(5)
Aggregate Affiliated Students (in hundreds)	(45)
Affiliated Enrollees	(3)
Affiliated Enrollees (in hundreds)	(43)
Aggregate Affiliated Enrollees	(6)
Aggregate Affiliated Enrollees (in hundreds)	(46)
Enrollee Load	(4)
Enrollee Load (in hundreds)	(44)
Aggregate Enrollee Load	(7)
Aggregate Enrollee Load (in hundreds)	(47)
Staff-Related Functional Bases	
Number of Academic Staff	(8)
Aggregate Number of Academic Staff	(12)
Number of Academic Support Staff	(9)
Aggregate Number of Academic Support Staff	(13)
Number of Nonacademic Staff	(10)
Aggregate Number of Nonacademic Staff	(14)
Total Staff	(11)
Aggregate Total Staff	(15)
Total Full-Time Academic Staff Hired	(35)
Number of Full-Time Staff	(50)
Cost-Related Functional Bases	
Total Academic Staff Salaries (in thousands)	(32)
Aggregate Total Academic Staff Salaries (in thousands)	(37)

Table 3 (Cont'd)

Cost-Related Functional Bases	
Total Academic Support Staff Salaries (in thousands)	(33)
Aggregate Total Academic Support Staff Salaries (in thousands)	(38)
Total Nonacademic Staff Salaries (in thousands)	(34)
Aggregate Total Nonacademic Staff Salaries (in thousands)	(39)
Total Staff Salaries (in thousands)	(36)
Aggregate Total Salaries (in thousands)	(40)
Operating Costs (in thousands)	(24)
Aggregate Operating Costs (in thousands)	(25)
Space-Related Functional Bases	
Classroom Space (in square feet)	(18)
Aggregate Classroom Space (in square feet)	(21)
Instructional plus Special Laboratory Space (in square feet)	(19)
Aggregate Instructional plus Special Laboratory Space (in square feet)	(22)
Total Space (in square feet)	(20)
Aggregate Total Space (in square feet)	(23)
Number of Stations (in a room)	(48)
Number of Square Feet (in a room)	(49)
Program/Curriculum-Related Functional Bases	
Number of Affiliated Programs	(16)
Aggregate Number of Affiliated Programs	(17)
Number of Directly Affiliated Cost Centers	(26)

Notes: Each functional base should include the words: "per cost center."

Figures in parentheses refer to a code needed by CAMPUS model to recognize particular functional bases.

FIGURE 2

The Command Structure in Campus

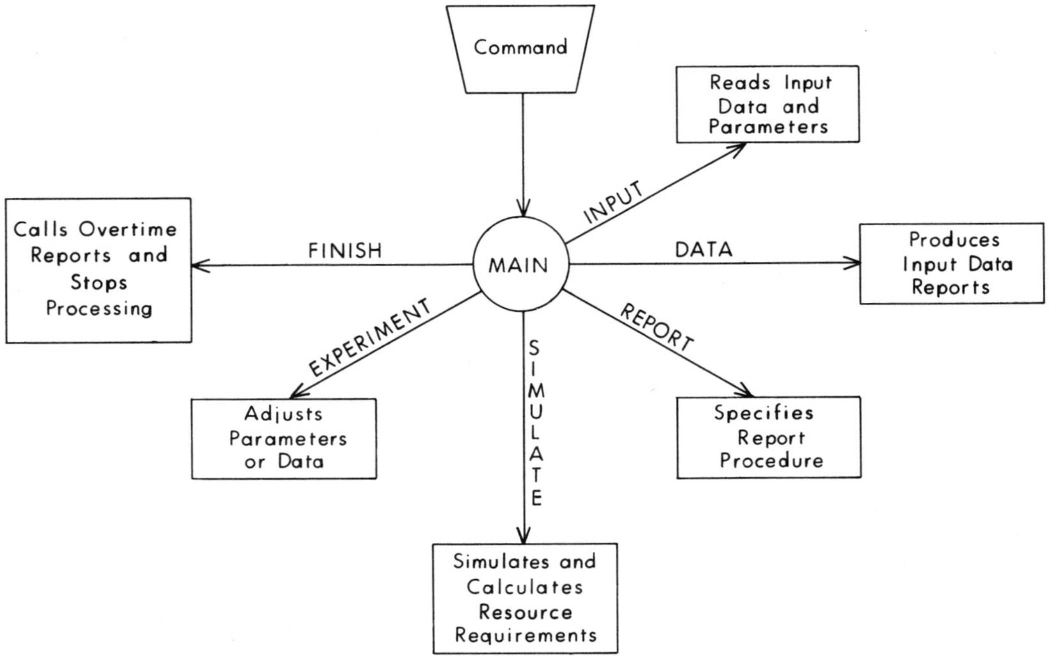

Source: Sceviour (1969), 19.

TABLE 4

Level 1 Command INPUT,
Index of Level 2 and Level 3 Command INPUT Documents

LEVEL 2	LEVEL 3	
DEFINE	01	Institution Name and Simulation Time Factors
	02	Cost Center Levels
	03	Cost Centers
	04	Program Levels
	05	Programs
	06	Program to Cost Center Affiliation
ACTIVITY	01	Activity Type
	02	Specialty Type
	03	Schedule Range
	04	Section Size Range
	05	Resource Combinations
	06	Activities
	07	Exception Activities
	08	Exception Resources
PROGRAM	01	Program Curricula
	02	Curricula Activities and Participation Rates
	03	Program Duration and Enrollment Update
	04	Credits per Credit Range by Program
STUDENT	01	New Entrants to Institution with NO Academic Credit
	02	Distribution of New Entrants with NO Academic Credit
	03	New Entrants with Academic Standing
	04	Student Transitions
	05	Student Credit Load
STAFF	01	Academic Staff Ranks
	02	Academic Staff Activity Teaching Duties
	03	Academic Staff Activity Nonteaching Duties
	04	Academic Support Staff
	05	Nonacademic staff
X-STAFF	01	Detailed Academic Staff Ranks
	02	Detailed Academic Staff Activity Teaching Duties
	03	Detailed Academic Staff Activity Nonteaching Duties

LEVEL 2	LEVEL 3	
	04	Detailed Academic Staff Inventory, Transition, and Hiring Criteria
	05	Detailed Academic Staff Optimization and Update Policies
SPACE	01	Classroom Sizes (stations)
	02	Laboratory Sizes (stations)
	03	Classroom Space Planning Factors
	04	Classroom Type Characteristics
	05	Instructional Lab. Space Planning Factors
	06	Instructional Lab. Type Characteristics
	07	Special Lab. Space Planning Factors
	08	Special Lab. Type Characteristics
	09	Service Space Characteristics by Type
	10	Cost Center Space Characteristics
	11	Service Code Specifications
	12	Space Category Codes, Names, Construction, and Maintenance Costs
	13	Miscellaneous Space Specifications
	14	Teaching Space Control Centers
AVLSPACE	01	Available Classroom Space
	02	Available Instructional Laboratory Space
	03	Available Instructional Special Laboratory Space
	04	Available Space by Category
SERVICE	01	Service Departments and Affiliations
	02	Service Staff
	03	Service Space
	04	Service Equipment
EQUIPMEN	01	Equipment Resource Characteristics
REVENUE	01	Characteristics of Revenue
	02	Revenue at Cost Centers
	03	Revenue at Programs
	04	Revenue of Service Departments
MISCELLA	01	Miscellaneous Resource Characteristics
	02	Miscellaneous Resource by Cost Center
INREPRT	01	Input Report Controls
	02	Comments
UTREPR	01	Output Report Control—Cost Centers
	02	Output Report Control—Program
OVTIME	01	Cost Center Output Reports Overtime

Space and Avlspace. There are four types of space in this section: classroom, instructional labs, special labs, and service department. Operating costs, construction costs, and service characteristics (e.g., air-conditioning) are also needed.

Service. Inputs are needed for staff, space, cost, and equipment associated with service departments.

Equipment. Information is required on cost and type of teaching equipment.

Revenue. Revenue must be specified by source and use.

Miscella. Forms must be provided for developing miscellaneous resources (e.g., benefits, travel expenses).

Inreprt. This command enables the user to call for "input data" reports.

Outrepr. This command enables the user to call for "cost center" reports.

Ovtime. This command enables the user to call for "overtime" reports (i.e., reports covering three quarters or two semesters).

The CAMPUS Process

Since CAMPUS is a comprehensive resource analysis model, it must consider the multiple missions of institutions of higher education. The program structure shown in Figure 2 is designed to classify the primary missions or objectives of higher education (instruction, research, public service) and also to classify the support missions (academic support, student support, institution support).[6]

The Instruction Programs

The instructional process in CAMPUS is one of the model's key strengths because it "forces" the user to define his degree-curriculum-course relationship. Before exploring the instructional process we need to expand slightly on the list of definitions.

Credit range is the level of academic standing. A typical four-year degree program has four credit ranges: freshman, sophomore, junior, senior.

Participation rate is the probability of taking a course.

Curriculum is a set of activities (courses) and their corresponding participation rates.

Chart 8 is a schematic representing the CAMPUS instructional process. Starting from the left side of the schematic the process begins with each student's being placed in a group element (a degree major). Associated with each degree (program element) are a varying number of curricula. One curriculum is required for each simulation period/credit range combination. A typical four-year degree would contain twelve curricula—three simulation periods (quarters) times four credit ranges.

Continuing to move from left to right on Chart 8, we note that each curriculum consists of a number of activities, each with an assigned participation rate. As noted previously, each activity is associated with and draws resources from an academic department (cost center).

To improve the reader's understanding of the instructional process, Table 5 presents some sample data. It is one of the "input data" reports available from CAMPUS. First, we note that this report is for the University of Minnesota School of Business, 1969-70. Then, starting from the left, we note that this is "program node" 13, MBA, day degree. The first curriculum shown, curriculum number 160, is for credit range 1 (first-year MBA) and simulation period 1 (fall quarter); it includes fifteen courses. Each activity or course has an assigned participation rate (e.g., activity no. 159, 75 percent) representing the probability of a first-year MBA student's taking this course in the fall quarter (1969-70). Continuing down Table 5, we note that curriculum number 162 (first-year MBA, winter quarter) includes twelve courses. Since an MBA degree is a two-academic-year program, the program element "MBA day" has six curricula (two credit ranges x three simulation periods).

The Support Programs

Developing resources for the support programs involves using the concept of a service department. A typical example of a service department is the computer center. To determine the "supplies" (one resource type) needed for this department it is necessary to use the functional basis concept. The fifty functional bases available in the model were shown in Table 3. A maximum of three functional bases can be used for each resource in each service department. The use of functional bases gives the model-builder a very flexible vehicle for determining resources in support programs.

CHART 8

Campus Instructional Process

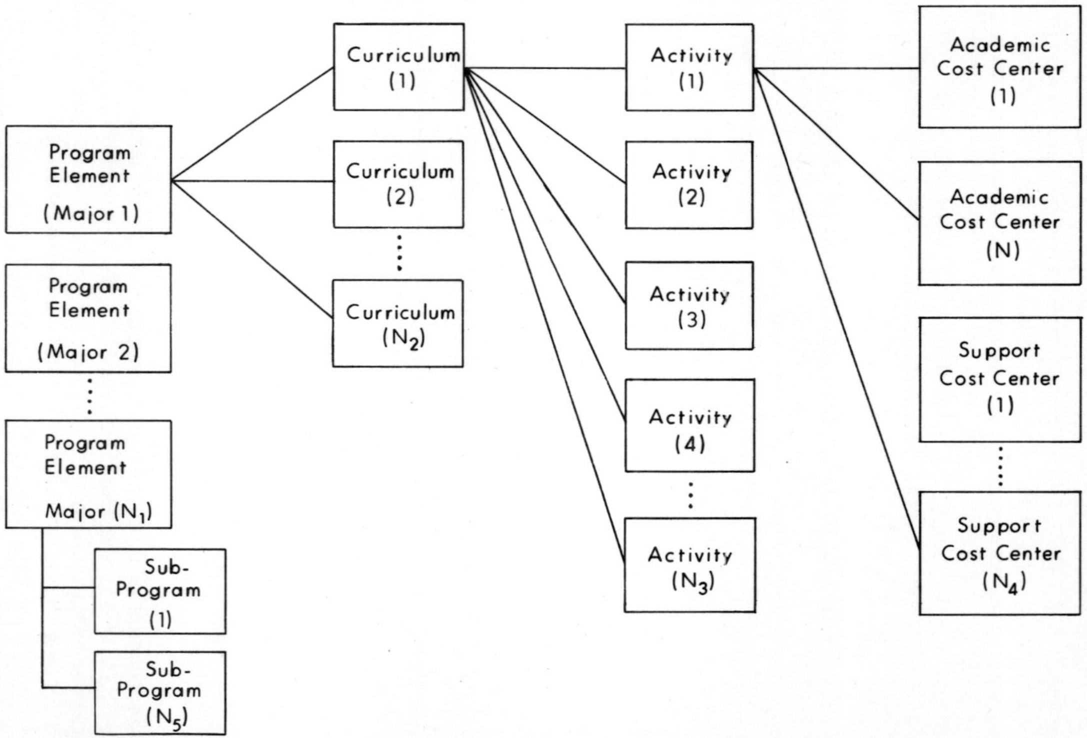

Note: Each activity has a corresponding participation rate.

TABLE 5

Input Data Report 3.1, CAMPUS Instruction Process,
University of Minnesota School of Business, Session 1969-70

Program Curricula and Activity Participation

Program Node	Program Name	Max. No. Years	Max. No. Credits	Credit Years	Simulation Period	Program Curriculum No. Code	Activity Number Code	Activity Calendar Code	Activity Type	Participation Rate (percent)
13	MBA Day	2	27	1	1	160	159	8006	Lecture 1	75
							161	8008	Lecture 1	75
							40	8260	Lecture 1	20
							65	8100	Lecture 1	10
							33	8050	Lecture 1	35
							243	8151	Lecture 1	35
							156	8001	Lecture 1	30
							64	8000	Lecture 1	10
							204	8000	Lecture 1	35
							95	5032	Lecture 1	10
							302	PSY1	Lecture 1	10
							300	E001	Lecture 1	10
							83	8158	Lecture 1	10
							137	8100	Lecture 1	30
							268	8154	Lecture 1	30
				1	2	162	162	8009	Lecture 1	75
							163	8010	Lecture 1	10
							262	3064	Lecture 1	10
							206	8047	Lecture 1	20
							65	8100	Lecture 1	20
							34	8051	Lecture 1	35
							244	8158	Lecture 1	35
							228	8150	Lecture 1	30
							95	5032	Lecture 1	10
							304	5001	Lecture 1	20
							301	E002	Lecture 1	20
							83	8158	Lecture 1	20
							302	PSY1	Lecture 1	20
							300	E001	Lecture 1	20

(Cont'd)

TABLE 5 (Cont'd)

			Program Curricula and Activity Participation						
Program Node	Program Name	Max. No. Years Credits	Credit Years	Simulation Period	Program Curriculum No. Code	Activity Number Code	Activity Calendar Code	Activity Type	Participation Rate (percent)
			1	3	164	252	8159	Lecture 1	75
						163	8010	Lecture 1	10
						226	5100	Lecture 1	20
						138	8101	Lecture 1	20
						262	3064	Lecture 1	20
						99	5012	Lecture 1	20
						95	5032	Lecture 1	20
						298	MATH	Lecture 1	20
						83	8158	Lecture 1	20
						64	8000	Lecture 1	25
						301	EC02	Lecture 1	20
			2	1	161	322	PL 3	Lecture 3	10
						323	PL 3	Lecture 3	10
						324	PL 3	Lecture 3	15
						325	PL 3	Lecture 3	10
						326	PL 3	Lecture 3	10
						327	PL 3	Lecture 3	15
						159	8006	Lecture 1	25
						161	8008	Lecture 1	25
						40	8260	Lecture 1	20
						65	8100	Lecture 1	10
						284	8053	Lecture 1	100
						302	PSY1	Lecture 1	20
						300	EC01	Lecture 1	20
			1	3	164	252	8159	Lecture 1	75
						163	8010	Lecture 1	10
						226	5100	Lecture 1	20
						138	8101	Lecture 1	20
						262	3064	Lecture 1	20
						99	5012	Lecture 1	20
						95	5032	Lecture 1	20
						298	MATH	Lecture 1	20
						83	8158	Lecture 1	20
						64	8000	Lecture 1	25
						301	EC02	Lecture 1	20

Program Node	Program Name	Max. No. Years Credits	Credit Years	Simulation Period	Program Curriculum No. Code	Activity Number Code	Activity Calendar Code	Activity Type	Participation Rate (percent)
			2	1	161	322	PL 3	Lecture 3	10
						323	PL 3	Lecture 3	10
						324	PL 3	Lecture 3	15
						325	PL 3	Lecture 3	10
						326	PL 3	Lecture 3	10
						327	PL 3	Lecture 3	15
						159	8006	Lecture 1	25
						161	8008	Lecture 1	25
						40	8260	Lecture 1	20
						65	8100	Lecture 1	10
						284	6053	Lecture 1	100
			2	2	163	322	PL 3	Lecture 3	10
						323	PL 3	Lecture 3	10
						324	PL 3	Lecture 3	15
						325	PL 3	Lecture 3	10
						326	PL 3	Lecture 3	40
						327	PL 3	Lecture 3	15
						162	8009	Lecture 1	25
						163	8010	Lecture 1	50
						262	3064	Lecture 1	10
						206	8047	Lecture 1	20
						65	8100	Lecture 1	10
						285	6053	Lecture 1	50
						285	6053	Lecture 1	50
			2	3	165	322	PL 3	Lecture 3	10
						323	PL 3	Lecture 3	10
						324	PL 3	Lecture 3	15
						325	PL 3	Lecture 3	10
						326	PL 3	Lecture 3	40
						327	PL 3	Lecture 3	15
						252	8159	Lecture 1	25
						163	8010	Lecture 1	30
						226	5100	Lecture 1	10
						138	8101	Lecture 1	20
						262	3064	Lecture 1	10
						99	5012	Lecture 1	20
						286	6053	Lecture 1	50

The Research and Public Service Programs

If the research or public service activity is carried on by a specified organization (e.g., MIS research center), determination of required resources can be handled analogously to those for support programs. If the research or public service is a "faculty activity," CAMPUS develops the required resources using a nonteaching duty category. Two possibilities are available: using a fixed percentage of faculty time or using the "functional basis" on a resource called "staffing units."

CAMPUS Outputs

There are three major groupings of output reports available in the present version of CAMPUS: input data reports, (2) cost center reports, and (3) overtime reports. The first group of reports is provided to collate the input information and develop it in a logical report format, thus facilitating an examination of the input data. There are forty-four input data reports.

The second group of reports, the cost center reports, are provided to aid the institution's managers, e.g., department heads, deans, etc. The present version of CAMPUS has seven major report types and forty-eight report formats.

The third group of available reports is overtime reports. These reports are for a session (year), in contrast with the two preceding groups, which are "single-period (quarter)" reports. Up to ten sessions are possible. Five categories of reports are available.

THE INSTITUTIONAL SETTING OF PROJECT PRIME

Given these technical aspects of CAMPUS, how well did it fit the needs of individual schools? Research for Project PRIME was coordinated by the Minnesota Higher Education Coordinating Commission. It has the responsibility for planning and coordinating the five systems of higher education in the state: the state college system, the junior college system, the vocational-technical schools, private colleges, and the University of Minnesota. The commission does not have the authority to review the budgets of the five systems, but it does have "program review" authority (granted by the 1971 legislature).

The 1971 legislature also changed the makeup of the commission from representatives of the five systems to a complete "layman" board. The commission's staff consists of approximately twelve professional personnel, primarily with backgrounds in education.

The PRIME project team consisted of a total of ten persons (some part-time) with a $137,000 budget for the one-year duration of the study.

Five institutions were involved in the test implementation. Each implementation followed a five-step procedure:

1. Orientation of key administrative and faculty personnel
2. Data collection and development of key policy parameters
3. Test run and debugging
4. Experimental runs
5. Documentation and review of results.

State College System

Initially the plan was to do only the behavioral science division of Bemidji State College, because of model restraints and the amount of data collection involved. Approximately two-thirds of the way through the project, the total school was put on the model. This was accomplished by aggregating courses, e.g., lower-division English, upper-division music, etc. Key persons at Bemidji include the president, who was quite negative at first but became a proponent; the head of the behavioral science department; and three people from the institutional research department. Bemidji ran three experiments and is still utilizing the model.

Junior College System

Initially the plan was to do only Lakewood Junior College, but after seeing the initial results, Rochester Junior College was added. Plans are in progress to collect data for all eighteen junior colleges. Key participants in this implementation were the president of Lakewood, the assistant chancellor for information systems, and the chancellor of the junior college system.

University of Minnesota, School of Business Administration

The initial data collection for the University of Minnesota, School of Business Administration, was done by the project team. Since the business school is both a graduate and an undergraduate institution, it was the most complete utilization of the model's capability. In conjunction with personnel from the Long-Range Planning Committee, twelve ten-year experiments were conducted, with various alternative assumptions about the administrative structure, degree offerings, and student flow. A planning "calendar" compatible with the present university budgeting cycle was proposed.

University of Minnesota, Division of Educational Psychology

The educational psychology implementation, not included in the original proposal, was attempted at the urging of the division head. He had the assistance of two graduate students, plus support from a professor on a year's leave. The latter was on leave to conduct a detailed review of the curriculum of the College of Education (of which educational psychology is a division). Plans are being laid to add the educational administration division. Although educational psychology did not conduct any experiments, considerable time was spent analyzing the significance and impact of the key input parameters.

Hopkins School District

Although it was not included in the original proposal, project personnel worked with the Hopkins school district. After determining that the model would simulate a complete school district, a three-year simulation was run on one high school. At this time, the data have not been used by the district's personnel, but they have expressed interest in continued experimentation with the model.

MODEL IMPROVEMENTS AND RELATED RESEARCH

Although the test implementations and the conversion of the program from IBM to CDC equipment were the major thrust of the PRIME project, several other related activities were conducted.

Documentation

Three categories of documentation were added to the documentation received from the Ford Foundation project at Toronto: (1) a user input manual, available as Project PRIME Report No. 12; (2) technical documentation, primarily comments in the computer code; and (3) user experimentation manual, documentation on how administrators can use the model, available as Project PRIME Report No. 9.

WICHE Compatibility

The PMS division of WICHE has had a significant impact on standardizing various codes referring to course numbering and student levels. Capability was added to the model to "label" each course and student with appropriate WICHE codes.

Subprograms

In order to improve the model's ability to predict class and section sizes, the project introduced the concept of a "subprogram." Subprograms are basically students who "minor" in certain fields.

Faculty Activities

The percent of time spent by faculty members in various activities is an input to the model. In order to determine these faculty inputs the project conducted a self-work sampling study. Self-work sampling studies have been done with engineers, with doctors, with clerical pools, and with medical students. A data card was provided to thirty-three faculty members on which they noted the amount of time spent on forty-three categories of activity, e.g., classroom contact, research and scholarly activity, etc. Cooperation from the faculty and administrators was excellent throughout the study.

Program Costing

The descriptive material from the University of Toronto on the CAMPUS model indicated that the computer code was capable of computing cost for both programs and cost centers. However, the computer code which was released was incapable of computing program costs. The only reports available were "cost center"-oriented, as explained previously.

In order to convert input-oriented department data into output-oriented program data, the project team designed a "program costing module." The reports available from the program costing module enable an administrator to develop and analyze in considerable detail his "program budget."

Four processes are used by the module to handle the conversion:

1. Service department process: a set of procedures to handle the staff, space, and equipment

2. Activity/curriculum process: a conversion routine to handle the six types of "direct cost" resources—academic staff; academic support staff; classroom maintenance cost; lab space maintenance cost; special Lab space maintenance and equipment operating costs; and teaching equipment cost

3. Nonteaching duty process: a set of rules for converting faculty nonteaching duties to program elements

4. Academic indirect resources process: an allocation technique for three types of "academic" indirect resources—nonacademic support (e.g., secretaries), miscellaneous resources (e.g., supplies), and office space maintenance cost.

The individual application of these four processes to the CAMPUS V model results in a series of program-oriented reports. If all four processes are applied, a series of summary reports (7.1-7.3) is available. A sample format for each of the available program reports can be found in Project PRIME Report No. 10. Table 6 includes the name of each report.

For each program element, it is possible to receive many of the above reports for ten years, by quarter. Typically a manager would not want to look at this number of reports. To correct this situation, a series of "program" overtime reports was developed. These reports summarize various operating costs, by year, for a ten-year period.

Available Reports

Table 6 contains an annotated list of the PRIME reports prepared by the project. Depending on the interest of the reader, several combinations of reports are relevant:

Potential User of CAMPUS: Reports 2, 15, and 16

Interested in Project PRIME: Reports 1, 2, 14, and 16

User of CAMPUS: Reports 8 and 12

PPB Systems in Higher Education: Reports 3, 4, and 10

Faculty Activity Analysis: Reports 6 and 7

TABLE 6

Project PRIME Reports: Annotated Bibliography

Report Number	Title	Author	Annotation
1	Test Implementation of CAMPUS (a Computer-Based Simulation Model) for Higher Education Administration and Planning in Minnesota. March, 1970. 59 pp.	Andrew, Cordes, Lorents	Initial proposal.
2	An Introduction to Project PRIME and CAMPUS-MINNESOTA. November 17, 1970. 31 pp.	Cordes	Describes in general CAMPUS inputs, processes, and outputs. Also provides overview of Project PRIME and research in progress.
3	PPBS in Higher Education: An Annotated Bibliography. May, 1971. 248 pp.	Cordes	Includes 325 annotated references; 100 carefully selected annotated references from other sources; 50 author's abstracts, 50 related bibliographies; and approximately 1,000 nonannotated references.
4	PPBS in Education: Concept, Operation, Status, and a School of Business Administration Example. March, 1970. 61 pp.	Cordes	Establishes six conceptual processes to describe PPB systems. Using these six processes as a framework, the Department of Defense's PPB system is then described in detail. After discussing the status of PPB in higher education, an approach to applying the system at a school of business is explored.
5	Program Costing with the CAMPUS Simulation Model. June, 1971. 31 pp.	Cordes	Program costing is the conversion of input-oriented budget data into output-oriented program data. The paper describes the philosophy of program costing, the mechanics of conversion, and the resulting reports. A comparison of a course-oriented and a degree-oriented program structure is also presented. (Should be read in conjunction with PRIME Report No. 8.)
6	Faculty Activity Analysis and Planning Models in Higher Education. June, 1971. 31 pp.	Lorents	Activity analysis and workload measurement are examined in the context of higher educational institutions.
7	A Faculty Activity Information Subsystem and CAMPUS-MINNESOTA. June, 1971. 13 pp.	Lorents	Develops a university information system that serves both transaction processing needs and information needs of management.
8	An Operational Overview of the CAMPUS Simulation. June, 1971. 48 pp.	Cordes	Describes in considerable detail a one-year project aimed at implementing the CAMPUS model in a statewide educational planning environment.
9	Using a Planning Model in Higher Education. (In progress.)	Fisher	Deals with operational aspects of planning models at various decision levels in higher education.

(Cont'd)

TABLE 6 (cont'd)

Report Number	Title	Author	Annotation
10	Resource Analysis Models in Higher Education: A Synthesis. (In progress.)	Cordes	A Ph. D. dissertation describing an approach to resource analysis for higher education. First, an integrated management framework is developed, based primarily on the ideas associated with PPB systems (see PRIME 4). Within this framework, the author presents a scheme for doing resource analysis based on three models: (1) CAMPUS; (2) program costing; and (3) input-output (each of these models is described in other PRIME reports). Three representative problems are also presented: (1) limiting admissions for an undergraduate program; (2) establishing one master's-level degree, instead of the present three degrees; and (3) reorganizing from a departmental structure with six academic departments and three research centers to a program organization with four research centers.
11	Converting CAMPUS V to CAMPUS-MINNESOTA. (In progress.)	Davitt	Describes the process of converting CAMPUS-V from an IBM 360-85 to a CDC 6600.
12	CAMPUS-MINNESOTA User Information Manual. June, 1971.	Andrew	The report is designed to aid the user in four specific areas: (1) the conceptual modeling of the institution; (2) the preparation of machine-readable input data; (3) the preparation of simulation and report commands for the model; (4) the actual running of the program on a CDC 6600 computer. Copies of the seventy-five input formats are included.
13	Applying Input/Output Analysis and the EL FYD Model to Higher Education. (In progress.)	Cordes	Describes the use of input/output analysis in higher education. The application represents the third model in the resource analysis schema described in PRIME 10. PRIME 8 and 5 should be read before this report.
14	Mid-Year Progress Report. January, 1971. 10 pp.	Andrew, Cordes, Lorents	Includes a description of progress on the six original project objectives. Four additional goals that were added to the project are described. Detailed financial information on the project is also included.
15	Case Studies of Resource Simulation in Education (a High School; a Junior College; a State College and Two Schools of a Large University. (In progress.)	Cordes (editor)	Each case study is written by an individual within the institution where the test implementation was attempted. Included are descriptions of data collection problems, usefulness of the model, and overall impressions of the test implementation.
16	Final Report of Project PRIME. (In progress.)	Andrew, Cordes, Lorents	Includes a summary of the project and conclusions and recommendations about using CAMPUS in Minnesota education.

Resource Analysis and Modeling: Reports 5, 8, 10, and 13

Experimentation with CAMPUS: Reports 9 and 10.

CONCLUSION: SOME PROS AND CONS OF USING A SIMULATING MODEL FOR IMPROVING PLANNING IN HIGHER EDUCATION

What next for CAMPUS in Minnesota education? That is the difficult question faced by the state's decision makers. Does it improve planning? Does it help structure information? Will it enable the higher education system to move more rapidly to a PPB system? To help answer these questions, this section explores the pros and cons of utilizing the CAMPUS model. Although the discussion is based on the experiences with one model in one state system, the reader may be able to generalize the comments and utilize them in his situation.

Cons

The model is restricted to institutions with no more than twenty-five cost centers (departments), eighty programs (degrees + support programs), and 1,000 activities (courses). The model was unable to handle Bemidji State College (5,000 enrollment) to the "course level" of detail. However, with proper "manipulation," the model was unable to simulate the total school to an "aggregate level" of detail.

The amount of data collection is significant, if an automated data base is not available. The amount and type of data required for each course was discussed previously. A second example of the magnitude of the data collection task is as it relates to the instructional process. Referring again to Chart 8, we see the program element-curriculum-activity relationship. If the program element is a four-year degree (e.g., Bachelor of Science, Business [B.S.B.]), there are twelve curricula (three quarters x four years). For each of the twelve curricula, a student in B.S.B. may take any of, say, fifteen courses. Associated with each of these fifteen courses is a probability (a participation rate) that B.S.B. students will elect to take this course. Thus, it is necessary to determine 180 probabilities per degree (twelve quarters x fifteen courses). For a school with 100 degrees, 18,000 participation rates are required. If the degree requirements change over the period of the simulation, the participation rates will also change.

The model is expensive to run. To run a ten-year simulation of the School of Business on the CDC 6600 costs over $100.

Pros

The model provides a framework for an MIS. As noted by Charles Nelson elsewhere in this book, "A major impediment to the development of management information systems is the weaknesses in the operating systems." He points out several examples of data needed in an MIS but not provided by the operating system programs. But without a model like CAMPUS, how does one know a priori what data are needed?

Developing the organization structure, program structure, and other policy parameters for CAMPUS is an invaluable experience. Overlapping courses, fuzzy degree requirements, and obsolete organizations become readily apparent.

A primary tool associated with PPB systems is a program and financial plan. Such a plan is a grouping of an organization's activities and resources into an output-oriented format and is PPB's data base. With the addition of the program costing module, CAMPUS has the ability to not only reproduce a program and financial plan but also to reproduce it for any number of alternatives.

Starting a PPB system in a higher education environment is difficult. A simulation model like CAMPUS can provide a convenient method of doing so.

The primary role of analysis can be seen as providing the decision-making process with systematic analysis concerning the organization's objectives, the alternative programs or activities available to accomplish these objectives, and the resources and benefit implication of each alternative. Simply stated, analysis in this context is needed continually to improve the plan. If the plan is a 500-page hard-cover document, it is difficult to keep it flexible or to analyze it. But if the plan is a CAMPUS tape, the analyst can experiment with various alternative configuration, present these alternatives to decision makers, and record the decision by updating the tape. In this manner direct and rapid interaction between the manager and the model is facilitated.

NOTES

1. For a description of each project, its objectives, and its status at the end of 1969 see Richard W. Judy, "A Research Progress Report on Systems Analysis for Efficient Resource Allocation in Higher Education" (University of Toronto, January 1, 1970). (Mimeographed.)

2. R. W. Judy, "Systems Analysis for Efficient Resource Allocation in Higher Education: A Report on the Development and Implementation of CAMPUS Techniques," in John Minter and Ben Lawrence, eds., Management Information Systems: Their Development and Use of the Administration of Higher Education (Boulder, Colorado: Western Interstate Commission on Higher Education, 1969), pp. 41-64.

3. Systems Research Group, The Development and Implementation of CAMPUS: A Computer-Based Planning and Budgeting and Information System for Universities and Colleges (Toronto: Systems Research Group, August, 1970).

4. For example, see the following works by Richard W. Judy: "A Research Progress Report on Systems Analysis for Efficient Resource Allocation in Higher Education" (Institute for the Quantitative Analysis of Social and Economic Policy, University of Toronto, January 1, 1970); (Mimeographed). "Costs: Theoretical and Methodological Issues," a paper presented at the North American Conference on Cost-Benefit Analysis of Manpower Policies, University of Wisconsin, Madison, May 14-15, 1969; and "Simulation and Rational Resource Allocation in Universities," in Efficiency in Resource Utilization (Paris: OECD, 1969), pp. 255-85. See also A. van Wijk, Using CAMPUS in Planning and Budgeting for Educational Institution (University of Toronto, 1970). (Mimeographed.)

5. R. W. Judy, J. B. Levine, and S. I. Centner, CAMPUS V Documentation, 1970, 6 Volumes (Toronto: Institute for Social Policy, 1970).

6. These six "programs" are part of the WICHE program classification structure. See Warren W. Gulko, Program Classification Structure (preliminary draft, June, 1970).

CHAPTER

7

AN ECONOMIST'S VIEW OF PLANNING PROBLEMS IN COMMUNITY COLLEGE SYSTEMS

Robert H. Fearn
Loren A. Ihnen

Community colleges are the fastest-growing sector of this country's postsecondary educational system. This chapter reviews the approaches to planning the regional and local needs for community colleges, and points to the possible ways that the control procedures established to monitor such institutions can provide incentives to move away from the original institutional plan. Ed.

INTRODUCTION

Educational research and planning activities have expanded rapidly since the 1950's, and many educational organizations are heavily involved in research and planning activities today. Educational organizations have always been involved in these activities. However, the relatively small-scale efforts in earlier periods were characterized by a somewhat haphazard and fragmented approach compared with the current deliberate and systematic approach.[1]

What forces have led to the recent concern and heavy involvement in educational research and planning? Surely many factors are involved, but three major ones seem to emerge. First, we are using many of our society's resources, both privately and publicly, to finance educational activities. Our current private and social affluence may be substantial relative to previous times or to other societies, but such a measure of affluence is misleading. The relevant comparison is to measure our wants and needs, both private and social, relative to the quantity of resources available to produce the goods and services required to satisfy these wants. Surely, most of us would agree that our resources are scare or limited relative to our wants. Educational activities must compete with other private and public activities for scarce resources; the competition has always been fierce and seems to be particularly difficult now.

A second factor is the role of education in the economic growth. There is considerable evidence that increases in education have contributed significantly to the rate of economic growth in the United States.[2] A very substantial research effort has been under-taken since the 1950's to specify and quantify more precisely the role of education in economic growth. Simultaneously, educational planning activities have expanded to increase efficiency within the educational process and to increase the contribution of educational activities to economic growth.

The third factor is the increased recognition that the quantity of education influences the individual's level of income and, hence, the distribution of income among individuals. The works by G. S. Becker, W. L. Hansen, G. Hanoch, and T. Johnson show that the rates of return to schooling, particularly at elementary and secondary levels, are large relative to estimated rates of return to investment in physical capital, and that the rate of return falls as the level of schooling increases.[3] J. Mincer and Becker have shown that investment in education and on-the-job training are important in explaining the distribution of income and income inequality.[4] Again this work has provided impetus both for further research on the effects of education and for additional planning to equate rates of return among alternative types of schooling and to reduce poverty and income inequality.

The last two factors cited, part of a body of long-standing literature in economics, have been rediscovered recently in "The Human Investment Revolution in Economic Thought."[5] The concept of investment in human knowledge and skills, i. e., human capital, provides us with an important theoretical framework for our discussion of educational planning.

THE CONCEPT OF HUMAN CAPITAL

Capital is a stock (of something) that yields a flow of services over time. This flow of services in future time periods may be used either directly in consumption or in the production of goods and services for consumption. Creation of this stock requires the use of resources in the current period, i. e., current consumption of goods and services which these resources could sustain is forgone. Thus, when we speak of investment in human capital we mean the creation of a stock of human knowledge, skills, etc. Education is not the only process through which investments in human capital can be made. Resources used for migration, improvements in health, on-the-job training, and improvements in labor market information may also represent investment in human capital; but this discussion is confined to educational investments. Not all educational activities are undertaken for investment purposes; some may be undertaken for immediate consumption or gratification. However, for purposes of illustration, educational activities are viewed primarily as investment activities.

Decisions regarding investment in education at the postsecondary level are made by the student or the student and his family. Educational institutions, which are producers and suppliers of educational services (screening, advising, teaching, testing, and certifying), and employers, who demand students' knowledge and skills, influence students' decisions. A student's decision is also influenced by other private individuals or groups and public agencies which affect either the costs of investing or the return from investing through grants, scholarships, subsidized loans, and taxation.

Therefore, the student can be viewed as a demander of educational services and as a supplier of labor and human capital services in the employment market. The student generally enters the labor market after completing schooling or training, but part-time students operate in both the educational market and the labor market simultaneously.[6] From the student's vantage point, the inputs involved in the educational production process are his time and effort, and school inputs, such as books, materials, building, equipment, teachers, etc. The cost of the student's time is his forgone earnings. The costs of the other inputs are his expenditures which are directly associated with undertaking schooling, e.g., tuition, fees, etc. All too often in discussions of the cost of investing in education only direct expenditures or costs are considered. However, forgone earnings generally are a larger share of the costs than are direct expenditures. The output of the educational process for the student is the increase in his knowledge and skills, which in turn leads to increases in his monetary and monetary income during his lifetime.

Students do not bear all costs or receive all benefits from investing in education. Thus, we must distinguish between private or student costs and returns and total or social costs and returns. Differences between the private account and the total account occur because of income transfers and external effects. By income transfers we mean subsidization of educational investment and taxation of the returns to that investment. Student payments in the form of tuition costs and fees at public educational institutions are only a small part of the average cost of providing the educational services. A. B. Carroll and L. A. Ihnen estimated the total subsidy for two years of postsecondary technical training at $1,458.[7] Hansen and B. A. Weisbrod estimated that public higher education subsidies in 1965 ranged from a total of $1,440 for two years at California junior colleges to $7,140 for four years at the University of California.[8] In addition, income forgone by students may be reduced by transfer payments in the form of scholarships, grants, G.I. Bill, etc. Not all of the added income from increased education accrues to the student investor. Income taxes reduce the income gain or return received and may result in income transfers to other members of society.

Educators have consistently argued that the benefits of increased education accrue not only to the student investor but also to other members of society now and in the future. These positive external effects are real social gains rather than income transfers among individuals, and these gains to other members of society must be added to the students' returns before taxes to obtain total or social benefits or returns. Often-cited benefits of additional education are increases in social awareness, political involvement, and other forms of good citizenship leading to a better society, which presumably benefits all its members. Unfortunately, efforts to identify and quantify the external benefits and costs of increased investment in education have been relatively unsuccessful to date.

THE EDUCATIONAL INSTITUTION AS A "FIRM"

Having described the student's role in educational and investment activities, we will now consider the role of educational institutions or organizations. Earlier these institutions were described as producers and suppliers of educational services. As decision-making units, both private and public educational institutions are concerned with functions similar to those performed by private firms or other public agencies. The educational institution purchases its resources (personnel, equipment, materials, etc.) in competition with other resource users. These resources must be organized and combined to produce an output. Finally, the output must be delivered and sold to the consumer.

To perform these functions effectively requires that the educational institution be concerned not only with its internal organization but also with its product market and its markets for inputs. Thus, not only is knowledge of its production function (the relationship between output and input) required, but information is needed about the supply function for these inputs and the demand function for its output. The supply function for teachers—that is, the relationship between salaries and number of teachers—is of special interest. The demand relationship is particularly complex because the students' demand for education is derived from the employers' demand for their knowledge and skills.

The educational institution is not a closed system operating in isolation from our economy and society. Instead, it is an open system operating in a national or, perhaps more realistically, an international economy and society.[9] This is not to suggest that it is inappropriate to study educational planning problems as a closed system. On the contrary, such an approach may be very fruitful for evaluating some short-run planning problems concerning internal efficiency of an educational organization. However, for intermediate and long-run planning purposes, it is more fruitful to view an educational organization as an open system obtaining resources from society and producing educational services, and hence additions to the stocks of human capital, which are returned to society. Activities undertaken in the educational system have an impact on the remainder of society, which in turn affects the educational system, and so on.

PLANNING FOR COMMUNITY COLLEGES

With the concept of investment in human capital in open educational and economic systems in mind, we will now consider more specifically the role of planning in the community college system.

One can begin by looking at four aspects of community college activities:

1. The various forces and participants who may move or seek to move the system in particular directions

2. The basic objectives of community college systems

3. The constraints faced by the systems

4. Certain questions about short-run and long-run planning.

Educational literature and several theories from the social sciences suggest the following forces and participants which may influence the direction of community college development:

1. Demands by students or prospective students for particular programs, given their tastes and their existing understanding of educational benefits and opportunities

2. Demands by local businesses and industries

3. Demands by local political authorities and/or local advisory groups

4. Internal incentives which may be implicit in the budgetary and accounting procedures of the system (or subsystem) in the presence of differential costs among the various curricula

5. The tastes and preferences of administrators of individual institutions with regard to types of educational programs

6. The plans promulgated by state or national political or educational administrative offices.

Although this discussion concentrates upon the central planning function, it is useful to delineate the other forces and participants—the other actors in the educational drama. We emphasize these other actors to keep our attention focused upon public education as a system which operates in an open environment. Each of the actors—particularly the students—has a set of options which he may choose (at some cost) and which will lead him toward his particular goal or goals. For the students, the options may include on-the-job training, apprenticeship training, private proprietary schools, military service training, and other approaches to investments in themselves.

THE OBJECTIVES OF COMMUNITY COLLEGES

Any person can list myriad objectives for a community college system, including general citizenship training, compensatory education, the development of specific salable skills in a local setting, and the like. As indicated in the introductory remarks, economists see education somewhat more broadly—specifically, as an investment process which develops human capital, from which flow streams of pecuniary and nonpecuniary benefits, largely in future time periods. For example, Herbert Parnes wrote:

> . . . the economic and cultural implications of education . . . are inextricably interrelated, and both must be taken into consideration in formulating educational policy . . . sound educational planning must rely on both these approaches for neither can logically be employed alone. The "cultural approach" is necessary because economic growth is not the only objective of societies, and no one would maintain that the sole function of education is to contribute to that end.[10]

How well does this general definition of education fit the community college subsystem? After working with community college problems for some time, it seems clear that "cultural" benefits do not play a predominant role in the community colleges and technical institutes. Administrators, teachers, and students seem primarily, although not exclusively, interested in the economic (or more precisely, the pecuniary) benefits which flow from education. Most of these benefits are "captured" by the students: some "spill over" to others in the society.

We need not emphasize the financial and legal constraints under which educational institutions operate. Whatever the specific form of the financial constraints, they represent—accurately or inaccurately—the underlying restrictions of scarce resources. Thus, it would appear that the operational objective of any educational system is to maximize the stream of pecuniary and nonpecuniary benefits flowing to the society, given the constraints imposed by scarce public and private resources at any point in time.

At the same time, it may be that private and public legal and financial constraints are so tight that individuals and society are forced to forgo opportunity gains; that is, the constraints may force the society to underinvest in education in general or in a particular subsystem or dimension. In this case it is the entire society which is not maximizing its returns, and reallocation of its scarce public resources toward (or away from) particular educational programs may yield higher social returns.

Based upon this definition of the objectives of education at the community college level, educational planning can be seen as a method for assaying current programs and resources and assessing the benefits and costs of alternative courses of action within the system. To the degree that the plans imply substantial social returns, both changes in the mix of current programs and changes in the magnitude of future programs may be considered. In the latter case, educational planners must be able to demonstrate that allocations of public funds for education will entail a higher social rate of return than alternative public investments or expenditures.

This view distinguishes between education planning and what can be called short-run operational guidance or management, a function which must be undertaken by the administrators in any system.

APPROACHES TO PLANNING

There are three major types of planning used to guide community college systems: the manpower requirements or "needs" approach, the social demand approach, and the benefit-cost or rate-of-return approach. Each approach has particular benefits and limitations.

The Manpower Requirements Approach

The manpower requirements approach has been the most widely used. North Carolina, for example, has used the manpower requirements approach, as have many other states and commonwealths.[11] In developing a manpower requirements plan, economists, manpower specialists, and educational planners use decennial census data and special surveys to ascertain previous and existing occupational-industry employment patterns and to forecast these patterns for future dates. An attempt is then made to determine what portion of the future requirements will be met by the stock of existing skills, migration, and other training programs. That number is subtracted from total requirements to find the "shortage" to be met in whole or in part by the public educational authorities.

Most forecasts are made by one of the following techniques:

1. Linear or curvilinear extrapolation of past employment patterns and trends into the future

2. Functional or factorized methods in which the future "needs" for trained manpower are seen as functions of relatively predictable variables (e. g., population or income)

3. Employers' forecasts of future manpower requirements.

The major analytical difficulty with the manpower requirements approach is that it represents, in essence, a "warehousing technique" that has little or no direct connection with the economics of labor markets and little recognition of the interconnections among the demand for and supply of different types of labor. The approach simply counts people and, as a general rule, abstracts both from (a) the influence of wages and employment conditions upon relative availabilities and demands and from (b) the substitutions in production of one occupational group for the other. The substitution possibility is known in the education literature as "job commonality" and is closely related to the concept of "job ladders." Economists generally refer to it as the elasticity of substitution.

Both of these limitations of the manpower requirements approach will be considered in turn. With regard to projections of future requirements, Adolf Sturmthal writes:

> The historic record does not show the demand for a given occupation, but rather the result of an interaction of demand and supply. The record is, therefore, not a very satisfactory basis for forecasting the demand to which future supplies are to be adjusted.[12]

When employers are surveyed, they are asked to declare their future "needs" without specifying expected wage or employment levels—to say nothing of expectations concerning trade union strength and practices. Employers therefore implicitly insert their own expectations. These, of course, can be expected to differ among respondents. In discussing employers' forecasts, Sturmthal finds that "expert opinion seems to agree that employers' estimates are reasonably reliable for short-run forecasts, but are of little use for forecasts extending further into the future, for—say—five to ten years."[13]

Next, we need to consider the difficulties caused by substitution possibilities. Because there are possibilities of substituting one worker for another with somewhat different skills via work force adjustment (specific retraining, job simplification, and the like), and because these possibilities grow over time, the manpower requirements approach becomes more inappropriate the longer the period for which it is used. In recent studies F. Welch and S. Bowles, working separately, found rather high elasticities of substitution among occupations or educational levels even in the short run.[14]

Sturmthal summarizes his and our critique of the manpower requirements approach, declaring: ". . . the simple truth is that at this time we do not possess a theory of the occupational structure of different industries which would enable us to make intelligent forecasts of the evolution of occupational structures."[15]

These largely negative comments are not intended to lead one to dismiss the manpower requirement approach or the need for forecasting. Indeed, efficient use of public resources implies explicit or implicit forecasting, and explicit methods are to be preferred. These comments are intended only to remind the producers and alert the consumers of the manpower requirements approach of its inherent limitations. The method cannot provide short-run operational guidance—with any reasonable degree of accuracy—nor does it provide a theoretically satisfying model for long-term planning. At best, it can provide a rough guide for the entire system (or subsystem) of education, outlining in broad dimensions the alternatives before the policy makers.

Within its inherent limitations, the manpower requirements approach can be an extremely useful device in helping the policy makers to consider the magnitude of their task and the trade-offs between broad categories of activities—e.g., adult basic, vocational, technical, and college parallel education.

The Social Demand Approach

The "social demand" approach is theoretically superior to the manpower requirements approach because it is based directly upon behavioral relationships.[16] Unfortunately, this approach is fraught with estimation problems. Although some progress is being made in solving these problems for education as a whole, this approach has yet to be applied to community college systems in a comprehensive way.

The Benefit-Cost Approach

The third planning approach is that of benefit-cost or rate of return analysis. The benefit-cost method involves a comparison of the present value of benefits with the present value of costs of the same program, using somc appropriate interest or discount rate to adjust future dollar values into present dollar equivalents. Alternatively, one may choose not to specify a discount rate, but to solve for the rate of return which is implicit in the benefit or cost streams—the so-called internal rate of return. Ideally, one would like to consider nonpecuniary benefits and costs together with the pecuniary ones. To the degree that one does not or cannot do so, and to the degree that nonpecuniary aspects are important determinants of behavior, the calculation of benefit-cost ratios or internal rates of return from pecuniary data represents partial analysis. Even where nonpecuniary benefits and costs are involved, benefit-cost or rate of return analysis can be used to advantage. If one is willing to specify that the net nonpecuniaries are positive or negative, the calculations based on pecuniary values can be regarded as minima or maxima. The limitation of rate of return analysis to pecuniary values (or to those aspects of the educational process which can be reasonably expressed in pecuniary form) does, however, caution against use of this method in a naive way when formulating educational plans.

In our view, the benefit-cost or rate of return approach is particularly well suited for the analysis of occupational education at community colleges, in that most of the gains appear to be pecuniary and most are "captured" by the individuals who have undergone the "treatment."

A few studies demonstrate some of the implications of this form of analysis. Carroll and Ihnen, for example, found a private rate of return to postsecondary technical education of 23.9 percent and a social rate of 16.7 percent.[17] These estimates reflect the most conservative of assumptions for earnings levels in the years beyond the range of available data. Moreover, no allowance was made for "cultural" benefits. By contrast, G. J. Stigler estimated the rate of return on capital invested in manufacturing industries to be 7.5 percent for the period 1939-56, and Becker estimated a private rate of return of 14.5 percent to college education by native white males among 1939 graduates and 13 percent among 1949 college graduates.[18] G. Hanoch and T. Johnson, using cross-sectional data from the 1960 census, also calculated marginal rates of return which generally are consistent with the patterns observed in other research work.[19]

Rate of return calculations such as those cited above can and should be made for both broad and narrow program groups in the community colleges. They provide considerable information and guidance to planners concerning the pecuniary "payoff" of the various programs. Note that this method of planning—unlike the manpower requirements approach—is based directly upon market valuation of public and private costs and benefits. To date too few studies have been completed to provide us with detailed estimates of these returns among major or detailed curricular programs or of the variation in returns for particular programs over time or between institutions. Nevertheless, the promise of rate of return analysis—as an analytically sound planning device for the community colleges—is clear.

The major failing of the benefit-cost or rate of return approach is that students must have completed various programs and have been employed for some time before reasonably reliable rates of return can be calculated. Thus, the information always lags behind the educational activity; it is useful for planning evaluations but useless—in and of itself—for projections. Nevertheless, the approach directs the attention of educational planners to benefits streams—giving them "equal time," so to speak, with costs. Such a concern with benefits induces follow-up systems for graduates of the community colleges in order to obtain information on income and earnings, job turnover, employer and employee satisfaction, and geographical, occupational, and industrial mobility. Indeed, these data may provide "early warnings" that particular programs or activities are not performing as expected. Poor showings on indicators such as low initial salaries relative to alternatives for untrained individuals of the same level of general ability, high turnover rates, and poor occupational matching—particularly when observed in combination—strongly suggest that private rates of return will not be as high as anticipated.

Furthermore, the apparent ex post nature of rate of return analysis does not totally inhibit its use in planning. Given certain assumptions concerning the sensitivity of wage levels to changes in supply and demand and given certain cost conditions, educational planners can calculate estimated rates of return for particular programs to use as guides for policy making. With adequate computational capacity, many such estimates can be made at relatively low cost. Use of rate of return analysis for evaluation and planning, however, requires that detailed information be obtained on benefits and costs among programs, institutes, and localities. Moreover, it requires that the benefit-cost relationships become an integral part of the short-run operational guidance system lest the short-run guidance procedures move the system in a contrary direction by overemphasizing costs relative to benefits.

LIMITATIONS AND "CURRICULUM DRIFT"

Although a consideration of the mechanisms for linking operational guidance procedures to the long-term plan is beyond the scope of this chapter, a brief discussion of the possible role of costs in directing the system seems to be warranted.

In a study for the Center for Occupational Education at North Carolina State University, R. M. Fearn demonstrated that regular budgetary and accounting procedures could create incentives to move the entire community college subsystem toward a mix of activities not envisioned in the overall plan.[20] In North Carolina, the allocation system for operating expenses for the present year is based upon full-time equivalent students (FTE's) enrolled in the previous year. The salaries of local administrative personnel are, in part, also tied to FTE's. Thus, where costs per student hour differ sharply among curricula, local administrators have an incentive to emphasize the low-variable-cost curricula at the expense of the high-cost curricula. E. B. Parry and E. F. Anderson found that costs per student for engineering technologies, health and medical curricula, and industrial and technical studies consistently exceed per-student costs for the liberal arts and college parallel training.[21] Parry, using North Carolina data for ten schools, found vocational and technical curricula to be more costly per membership (or contact) hour than college parallel curricula. Despite certain difficulties in the allocation of common costs discussed by Ihnen, these cost data, in conjunction with economic theory, suggest an incentive for what we call "curriculum drift" toward the low-variable-cost curricula.[22] Such a drift has occurred both within and between curriculum groups among the fifty-three community colleges and technical institutes in North Carolina.

The important point is that the incentive produced by this kind of "cost consciousness" exists without reference to future benefit streams. What is important in an investment plan is not the cost of the investment but the relationship of benefits to costs—the internal rate of return. Without an overall plan intended to reflect the relative size of benefit flows, there will be a strong tendency for costs to dominate the system. Indeed, administrators who are highly successful in "keeping costs down" may be lauded for their skills, even when their cost-consciousness leads them to expand low-rate-of-return educational programs at the expense of high-return programs. Thus, substantial opportunity losses may be imposed upon the society and upon the students who unknowingly expend their scarce resources of time and money in these areas of low productivity.

To summarize, each of the planning approaches available to community colleges has substantial limitations. Reasonable use of the available methods, however, can avoid gross errors and can bring the system closer to maximizing social benefits.

NOTES

1. M. Blaug, "Approaches to Economic Planning," Economic Journal, LXXVII, 306 (1967), 262-87.

2. E. F. Denison, "Measuring the Contribution of Education (and the Residual) to Economic Growth," in The Residual Factor and Economic Growth (Paris: OECD, 1964), pp. 13-55.

3. G. S. Becker, Human Capital: A Theoretical and Empirical Analysis, with Special Reference to Education (New York: Columbia University Press, 1964); W. L. Hansen, "Total and Private Rates of Return to Investment in Schooling," Journal of Political Economy, LXXI, 2 (1963), 128-41; G. Hanoch, "An Economic Analysis of Earnings and Schooling," Journal of Human Resources, II, 3 (1967), 310-29; T. Johnson, "Returns from Investment in Schooling and On-the-Job Training," unpublished Ph. D. dissertation (Raleigh, N.C.: Economics Department, North Carolina State University, 1969).

4. J. Mincer, "Investment in Human Capital and Income Distribution," Journal of Political Economy, LXVI, 4 (1958), 281-300; G. S. Becker, "Human Capital and the Personal Distribution of Income," W. S. Woytinsky lecture no. 1 (Ann Arbor, Mich.: Department of Economics, University of Michigan, 1967).

5. B. F. Kiker, "The Historical Roots of the Concept of Human Capital," Journal of Political Economy, LXXIV, 5 (1966), 481-99; M. J. Bowman, "The Human Investment Revolution in Economic Thought," Sociology of Education, XLIX, 2 (1966), 111-37.

6. R. M. Fearn, "Labor Force and School Participation of Teenagers," unpublished Ph. D. dissertation (Chicago: Department of Economics, University of Chicago, 1968).

7. A. B. Carroll and L. A. Ihnen, "Costs and Returns for Two Years of Postsecondary Technical Schooling: A Pilot Study," Journal of Political Economy, LXXV, 6 (1967), 862-73.

8. W. L. Hansen and B. A. Weisbrod, "The Distribution of Costs and Direct Benefits of Public Higher Education: The Case of California," Journal of Human Resources, IV, 2 (1969), 176-91

9. J. A. Thomas, "An Economic Approach to Systems Analyses," paper prepared for 1969 American Educational Research Association convention (Los Angeles, 1969).

10. H. S. Parnes, ed., Planning Education for Economic and Social Development (Paris: OECD, 1963).

11. A. A. Cruze, Strategic Plan for the North Carolina Community College System (Triangle Park, N.C.: Operations Research and Economics Division, Research Triangle Institute, 1970).

12. A. Sturmthal, "Current Manpower Problems, an Introductory Survey" (Urbana, Ill.: Institute of Labor and Industrial Relations, University of Illinois, 1964). (Mimeographed.)

13. Ibid.

14. F. Welch, "Linear Synthesis of Skill Distribution," Journal of Human Resources, IV, 3 (1969), 311-27; S. Bowles, "Aggregation of Labor Inputs in the Economics of Growth and Planning: Experiments with a Two-Level CES Function," Journal of Political Economy, LXXVIII, 1 (1970), 68-81.

15. Sturmthal, op. cit., p. 53.

16. Blaug, op. cit.

17. Carroll and Ihnen, op. cit.

18. G. J. Stigler, Capital and Rates of Return in Manufacturing Industries (Princeton, N.J.: Princeton University Press, 1965); Becker, Human Capital.

19. Hanoch, op. cit.; Johnson, op. cit.

20. R. M. Fearn, "Labor Markets, Incentives and Occupational Education," Center for Occupational Education Research Report no. 3 (Raleigh, N.C.: North Carolina State University, 1969). (Mimeographed.)

21. E. B. Parry, "An Investigation of Cost Differentials Between Trade, Technical, and College Parallel Curriculums" (Raleigh, N. C.: Department of Community Colleges, North Carolina Board of Higher Education 1968) (Mimeographed); E. F. Anderson, Differential Costs of Curricula in Comprehensive Junior Colleges (Urbana, Ill.: Bureau of Educational Research, University of Illinois, 1966).

22. L. A. Ihnen, "Vocational and Technical Education: Costs and Returns Studies," paper prepared for the 1969 American Educational Research Association convention (Los Angeles, 1969).

CHAPTER

8

DOES UNDERGRADUATE EDUCATION REALLY PAY OFF FOR SOCIETY?

Duncan Bailey
Charles Schotta

The personal return from investment in higher education, i.e., the enhancement of lifetime earnings of college graduates vs. nongraduates, is fairly well documented in economics literature. The question can be fairly put: Is there a return to the taxpayers who support public colleges and universities? Duncan Bailey and Charles Schotta explore these issues, using cost data from selected University of California campuses and employment statistics from the state of California. Ed.

INTRODUCTION

It is appropriately within the spirit of our contribution to the decision-making process in higher education that we begin this Chapter with what appears to be a fanciful anecdote from industry (which in the end is really a cruel joke directed at those of us who are responsible for discharging the task of spending the public's resources for higher education wisely).

There was once a man who was hired by a large factory, was taken by the plant manager to the assembly line, and there placed at a station with a number of bins filled with a bewildering array of parts. He was told to wait, and when the product came down the line, to put a part on it. He waited, and a rather amorphous object came rolling down the assembly line toward him. He picked up one part and could find no place to put it; he laid it aside. He picked up another part and again could find no place to put it; he laid it aside. After trying half a dozen times to find a part to fit the object that was being assembled, he returned in some frustration to the plant manager and said, "Look here, I don't know which of those parts I'm supposed to put on this thing. And incidentally, what are we making?"

Said the plant manager, a wry grin on his face, "I don't know what we're making. Use your own judgment about what part to put on it—but be sure to do a good job."

Those of us concerned with the ongoing operation of public institutions of higher education have a certain sympathy for the plight of both the factory manager and his new assembly-line worker. In this chapter, we address a problem faced by all decision makers in higher education. This problem is not unique to higher education. Governments, in operations other than the provision of higher education—e.g., defense, health and welfare, transportation, etc.—face very much the same problem as universities: how much of society's resources we should devote to the production of whatever public good our particular agency happens to produce. For the state-supported university, this means, first, what percentage of our state's tax revenues can justifiably be used to provide higher education; and second, recognizing that higher education is not a formless, unidentifiable "something" but, rather, a collection of identifiable curricula, each providing its own special interests and set of skills, what percentage of our social resource commitment shall be applied to each of the many curricula existing in each of our respective universities.

This total problem is an intriguing one and an important one, in view of the percentage of tax revenues committed to it today. The scope of this discussion is limited to an attempt to find an answer, i.e., some operational guidelines for decision makers facing the first problem. We seek to develop an economic rationale for the provision of social resources, through tax revenues, for the production of higher education.

In the past it was a sufficient argument to present to legislators and governmental decision makers, when requesting next year's budget for the university, that "higher education is a good thing"; but there are now many other social needs for which similar arguments can be made. The thoughtful state university official, in the face of rising public demand for economy in government and for tax reform, and other cries for governmental responsibility, must ask himself on what grounds an argument for a greater share of society's purse for the university can rest other than an emotional appeal for "education."

Several disclaimers are appropriate at the outset. First, nothing we have to say has any special relevance to privately funded institutions of higher education. We also wish to make very clear that this analysis considers as a benefit of higher education only the increase of total social product which occurs as a result of a better-educated citizenry. We completely neglect the benefits of education which Edward Bellamy described in his provocative Looking Backward when he made the case for universal education by asserting how much better it was to cultivate society until it became a rose garden rather than simply have one man hold a perfumed handkerchief to his nose to blot out the smell of the nauseating bog in which he resided. This neglect of the "consumption benefit" of higher education is a direct consequence of our view of the product of higher education which we characterize as "human capital," valued only in terms of the contribution it makes when entered into society's production function.

This chapter is divided into several major sections. First, the rationale for viewing the output of institutions of higher education as "human capital" is presented. While this discussion will be familiar to those who are steeped in recent economic literature on higher education, it may very well appear to some to be novel, to others radical, and to yet others downright wrong and dangerous. Nevertheless, acceptance of such a concept is fundamental to the application of the kinds of techniques which support requests for a larger share of society's resources for higher education.

After we have discussed the idea that what students possess when they leave the university is a piece of capital embodied in themselves, we then describe the data which show the basic difference between people who are working in society without benefit of the product of higher education and those who have secured this product.

A technique for measuring in a meaningful way the differences between the two basic income streams is then discussed. In order to use these highly suggestive present-value techniques, it is necessary to make certain assumptions to secure useful results. We outline and detail these assumptions as the analysis progresses and indicate the consequences of their relaxation.

Next, some estimates are presented of the private lifetime rate of return which individuals who have secured a college education enjoy, relative to those who have not. In order to compensate for the inability to look directly at the benefits to society in general which accrue because some receive higher education, an inference is made as to the magnitude of this benefit which people who do not have higher education must realize in order to justify social purchase of this portion of the individual's education. This is the question of the "spillover" benefits to society.

By the time that we complete this analysis, our conclusions will be clear and the procedures to be followed for making optimal decisions on resource allocation with respect to higher education will be apparent. We then go on to use the results of our analysis to look at several important allocation decisions faced by officials of higher education institutions.

The final section of the chapter consists of a summary of our results and an appraisal of their applicability to the resource planning activities of the higher education system of a state. It should be mentioned here that although all of our data pertain to the state of California, to the extent that the relationship between the program costs in universities and social product of other states is similar to that prevailing in California, our results are directly applicable to higher education systems in states other than California. It must be admitted that our results, though not our techniques, are less applicable to units of higher education that do not embody the strong research component contained in the University of California system.

THE HIGHER EDUCATION BUSINESS

It is fashionable today for representatives of higher education to make the case for a larger share of the public purse by claiming, as did the chancellor of the University of California at Irvine, that although the lifetime earnings potential of high school graduates is some $260, 000, the lifetime earnings of a college graduate averages around $400, 000—or a difference of about $140, 000.

In a similar vein, Charles Hitch, president of the University of California, referred to the other leg of the public education argument by claiming that a college graduate in California, over his entire working life, pays the state of California and local governments almost $9, 000 more in taxes than will a high school graduate. He further observed that this same college graduate pays a federal tax bill that is approximately $16, 000 greater, over his working life, than that paid by the average high school graduate.

What administrators and others who are concerned with higher education and its financing would like to show by such claims is that the increase in future social product through investment in education is as great as or greater than the increase in productivity that could be secured from alternative uses of current resources.

For the purpose of planning the input of social resources to the educational process, if one views this use of resources as an alternative to other uses of resources, the rate of return to resource use in education versus these other uses must be calculated. In order to do this, it is necessary to characterize the output of the educational process. Currently, two views of the fundamental character of the results of education exist.

The older view, and one which still dominates much educational planning, regards this output as essentially a consumer good. That is, the purpose of education is not to produce a skill in the student which can be used in a later production process but, rather, to produce the capacity to appreciate, in some larger sense, the world around him. Since whatever has been created by the educational process in the student is not a factor input, it must be valued in the same sense as other consumer goods which are not inputs to some production process.

The second view which can be taken of the output of educational institutions was first popularized by Theodore Schultz. Here the output of the educational process, i.e., the skill produced in the student, is valued for its contribution to the production process for which it is viewed as a resource input.[1] This view has considerable attraction for policy planning purposes, since it allows the planner to quantify much more easily the value of the output of the educational process. In this view, then, the education industry is simply a producer of capital goods—human capital or human skills.

The concept of human capital is not without theoretical difficulty. When one compares the skills possessed by, for example, a doctor with the output which results from the application to the production process of, say, a drill press, one is immediately struck by the fact that the drill press can be separated from the owner of this piece of capital and sold on the market. In the case of the doctor, the skill or human capital cannot be separated from its owner, nor can it be separated from the labor input which is embodied in the owner of the human capital. We can directly measure the value of the drill press by market sale price; but we must infer the present value of a piece of human capital by making some assumption about the value of the labor services of the individual, subtracting this from some discounted value of the stream of earnings in the future which will accrue to the individual who possesses both this labor skill and the embodied human capital. In spite of these conceptual difficulties, educational planning must proceed by analyzing the incremental investment rate of return to human capital.

For purposes of discussion, we are interested only in the investment aspect of the output of the education process; thus we disregard completely any consumption good output which the educational process produces. The major reason we are unable to value the consumption portion of educational output is that the person who acquires this consumption good (i.e., the student) usually pays nothing or very little for the good and, in contrast with the capital components, is never sold. By and large, tuition in most educational processes is much less than average cost.* Hence we cannot rely on market signals to value this consumption component.

There is another element in the total return to human capital which is very difficult to measure: the increase in productivity of labor, in general, and of other capital instruments which arises when additional human capital with higher skill levels is created. That the creation of human capital, through the educational process, confers significant externalities has long been recognized. However, the problem of measuring this exact return is little different from measuring the benefits of any product which gives off its services primarily in the form of the public good.[2]

In this analysis no attempt is made to measure directly the nonmarket social return. However, some conjecture is made as to the magnitude this return must assume for past levels of resource commitment to undergraduate education to have been socially rational.

LIFETIME INCOME STREAMS

The primary reason we felt it necessary to prepare our own data set lay in the inadequacies of the data used by previous researchers in this area. Gary Becker, in his admittedly seminal work on this topic, and later Bruce Wilkinson, both used census data on educational level and associated income.[3] For our purposes, the data, now nearly ten years old and highly aggregative, were sufficiently detailed to allow us to draw the conclusions necessary for practical educational investment planning.

We required a set of data which represents, for holders of undergraduate degrees in various occupations as well as for high school graduates, an accurate estimate of the current market value of particular skill levels at every stage in the working career. Fortunately the California State Personnel Board conducts an annual salary survey in which it queries private employers of a large number

---

*Some notable exceptions are for-profit educational enterprises such as flying schools, barber colleges, computer-programming schools, etc. Recent court decisions may cause more for-profit colleges to be established.

of occupational classifications concerning wages, hours, and conditions of work.[4] These data form the input to the California State Personnel Board's own formulation of wage rates for the various occupations required in the conduct of the daily business of a state as complex as California.

Working with members of the California State Personnel Board, we were able to construct the expected annual income, year by year, which the average employee in seventeen different college and non-college occupations might reasonably expect to receive. Thus we have, for a set of occupations requiring only a high school diploma, the entering salary for the first year of working life and the salary earned in each subsequent year through retirement (based on average expected promotion schedules). We have the same information for a set of occupations that require the college degree for entrance, from the first payroll year through the last.

Thus we have lifetime earnings, year by year, for seventeen occupations with either forty-seven or forty-three years of length.* These form the basic lifetime-earning-stream data for high school and undergraduate education. Though we will refer to this point again in our discussion of the basic assumptions of our study, we would like to emphasize here that these incomes are in constant 1966 dollars and hence do not reflect the forces of inflation.

INSTRUCTIONAL COSTS

What do these raw figures tell us? We could, of course, secure the differentials—college occupation less high school occupation—year by year, and sum them. Even if we do so, before we can ascertain whether the pecuniary returns realized from a college education are worthwhile, we must consider carefully what investment is necessary to reap these benefits.

Therefore the second part of our data collection effort was to secure the cost of producing a baccalaureate degree. Our basic cost estimates are for the University of California system for the 1966 fiscal year. It was not possible to secure a complete enumeration of the total cost of operating the teaching program. However, let us describe the cost elements included in our estimates.

---

*We neglect "fringe" benefits and retirement earnings in our calculation of earnings streams. The former may bias our results in an as yet undetermined fashion, while the latter drops out because of our use of present-value techniques.

All of our cost figures are derived from the current operating allowances appropriated by the California Assembly for the support of teaching programs of the nine University of California campuses. A general appropriation is made for the entire university for current operating expenses; and an allocation for each campus, as well as for the central administration located at the offices in Berkeley, is secured. The university requests its teaching program allocation on the basis of the number of full-time equivalent (FTE) students which it expects for the coming fiscal year. The cost allocation formula used for the University of California is such that all lower-division undergraduate FTE's are entered with a weight of 1, all upper-division undergraduate FTE's with a weight of 1.5, all first-stage graduate and professional FTE's with a weight of 2.5, and second-stage graduate FTE's with a weight of 3. Thus, six second-stage graduate students are equivalent, for budget planning purposes, to eighteen undergraduate lower-division FTE's.

To determine the current operating costs of the University of California system chargeable to the various levels of students by campus, we first allocated the administrative overhead expense of the university to the various campuses on the basis of the percentage of the students in the total University of California system who were resident at each of the campuses.

Taking the total current operating expense appropriated by the legislature for 1966, we multiplied the total number of FTE lower-division, upper-division, first-stage graduate, second-stage graduate, and professional program students by their appropriate weights. We secured the percentage of total weighted FTE represented by each of these categories, and apportioned the total operating allowance of each campus to the various levels of students according to the percentage they represented of the total FTE of the campus. To secure average cost figures for each level, we merely divided these cost allocations by the actual physical number of FTE students in each of the categories.

Since some of the campuses are newly established and thus are experiencing high start-up costs, we did not calculate costs for any but the five well-established campuses: Berkeley, Los Angeles, Santa Barbara, Davis, and Riverside. The University of California at San Francisco (San Francisco Medical Center) was not considered because it is a professional school. For 1966 the average cost figures for these five campuses are presented in Table 7.

TABLE 7

Average Current Operating Cost per Student Level,
Selected University of California Campuses,
1965-66 Academic Year

Campus	Lower Division	Upper Division
UCLA	$1,873	$2,817
Riverside	$2,975	$4,470
Berkeley	$1,566	$2,358
Davis	$2,646	$3,977
Santa Barbara	$1,565	$2,346

Source: State of California, Support and Local Assistance Budget, 1966 Session (Sacramento, 1966).

These costs range from a low of $1,565-$1,566 average per year for lower division at Berkeley and Santa Barbara to a high of $2,975 for the lower-division year at Riverside. Upper-division undergraduate expenditures ran on the low side at Santa Barbara ($2,346) and Berkeley ($2,358) and on the high side at Riverside ($4,470).

While one could merely add up four years of expenses and call this the cost, i.e., the current operating cost to the taxpayer of maintaining a student for four years in the University of California, this clearly would not represent the cost of producing a B.A. degree for two reasons.

First, these cost figures contain only current operating expenses. It was not possible to estimate the capital consumption allowance of the charge which should be made for the use of all fixed university facilities which have been funded out of previous capital budgets. Second, many students remain for only one year, two years, or three years, with a relatively small percentage of those matriculating remaining to take degrees.

For the Berkeley campus we secured estimates of the annual exit rate.[5] As an example, for the class of 1964, 2,070 students had to be admitted to produce 1,000 graduating B.A.'s. Based on an allowance for intercampus transfers and on this attrition rate, we have been able to prepare an estimate of the average cost of one B.A. degree at the entire set of the University of California campuses which we have included in our study. It should be remembered that this figure is for current operating cost appropriated by the legislature and does not include the cost of using the physical facilities of the university.

These estimates are presented in Table 8. We estimate first the present-value cost of the B.A. at the University of California, Berkeley, without allowing for intercampus transfers, using both 5 percent and 8 percent discount rates. We then estimate the present-value cost of a B.A. at the five campuses we consider, allowing for intercampus transfers. The total five-campus average current operating cost of the B.A. at a 5 percent discount rate is $9,158.

Briefly, then, we have secured a lifetime income stream for seventeen occupations that require either high school graduation or a baccalaureate degree. We have also secured present values of the current operating costs of producing a B.A. degree in the University of California system.

ASSUMPTIONS AND TECHNIQUES

There are several basic assumptions inherent in this type of study. The first is that we are dealing with constant dollar amounts, year by year, in order to eliminate the effect of inflation. Second, in constructing time streams of income, it is assumed that there are

TABLE 8

Present Values of Operating Costs per B.A. Degree, 1966, Discounted at Two Rates

	Discounted at 5 Percent	Discounted at 8 Percent
Berkeley[a]	$9,303	$8,719
University of California[b]	$9,158	$8,555

[a]Intercampus transfers not included.
[b]Including intercampus transfers.

no relative changes in the structure of supply and demand in the labor market which would cause incomes of college-degree holders to rise faster than incomes of noncollege-degree holders, or vice versa. Third, it is assumed that the wage which is paid is identical to the value of the products produced by the worker at the margin—or, in economic terms, that the wage is equal to the marginal revenue product of labor. In our income streams for holders of the B. A. degree and holders of high school diplomas, we have incorporated the state tax structure of California, the federal income tax structure, and the local tax structures, so that these figures are net of tax. When we secure the present value of these time streams of income, we do so in a manner which reflects the structure of draft patterns, marriage patterns, and family formation patterns which prevailed in 1966.[6] We further assume that the so-called "fiscal dividend," the percentage of the federal taxes paid by inhabitants of a state that is returned to the state in the form of services, is unity. (In general, nationally, this is seemingly reasonable, although there is some evidence that recently the fiscal dividend in California has been slightly in excess of 1.)

In order to convert all of these income streams and all of these cost streams into figures which are comparable, we have discounted each stream in order to secure the present value. The present value of the high school diploma is secured for an individual who surveys his lifetime stream at age eighteen, and the present value of the B. A. degree is also secured at age eighteen, the age at which most college-going decisions are made. These present-value streams are secured using discount rates of 5 percent and 8 percent.[7] It is assumed that 5 percent represents a relevant discount rate which society faces in considering public investments and that 8 percent represents the interest rate which a private individual should use in calculating the profitability of one investment relative to another. The cost figures have also been discounted to the present using a rate of 5 percent, since all cost figures are paid by the state.

FINDINGS

In this section we join our estimates of the present value of the differential income stream for college occupations over high school occupations with information on federal, local, and state tax rates and the cost of producing a B. A. degree at the University of California, together with a consideration of private cost which individuals incur, in order to estimate the private and social rate of return to undergraduate education.

Table 9 summarizes the present values of the income streams

and the present values of tax streams paid by individuals with college degrees and those with high school diplomas. As mentioned above, we use an 8 percent discount rate to secure the present value to the individual of the incremental income stream for having attended college, and a 5 percent discount rate to evaluate the present-value differential to the state. The first figures presented in Table 9 are present-value net incomes for college occupations, high school occupations, and the present-value differences of college over high school. This present-value figure is net of federal income taxes prevailing in 1966, the California income tax prevailing in 1966, the California sales tax prevailing in 1966, and an approximation of local taxes. Both income taxes are calculated on the basis of draft patterns, marriage patterns, and family formation patterns which were prevalent in 1966.

The figure for the college graduate of $71, 689 represents the present value to him of his lifetime net income evaluated at age eighteen at an 8 percent discount rate. The corresponding figure for the high school graduate of $66, 617 similarly represents the present value of his net lifetime income stream. Since there is a positive difference of $5, 072 between the present value of the college occupation and the present value of the high school occupation net of all taxes, the private rate of return to this investment on the part of the individual who chooses to attend college is quite clearly much greater than 8 percent.

Indeed, even if we required the entering college student to pay a fee of $5, 072, it might very well be that he would want to secure a college education simply for the consumption benefit that this would confer on him. Whatever the case, quite clearly, from an individual standpoint, so long as the expenses of attending college which would not be incurred if the person did not attend college sum to less than $5, 072, then in terms of an investment, if alternatives yield 8 percent or less, higher education is a "good buy."*

---

*Frequently it is asserted that the cost of maintaining a child in college amounts to around $2, 500 per year; thus one adds up the cost of going to college as some $10, 000. Clearly we cannot include the entire sum in assessing the investment rate of return to an individual for attending college, since the largest portion of this total outlay is for living expenses which would be incurred whether the person attended college or entered the labor force. The only element of private cost which can properly be charged to attending college are those outlays which would not have been incurred had the person not attended college: books, tuition, and the like. By evaluating the income stream at eighteen, and including four zero-income/earnings years while a person attends college, we have taken account of the

TABLE 9

Present Values of Income Streams

	Present Values at 5 Percent	Present Values at 8 Percent
Net Income		
College	---	$71,689
High School	---	66,617
Difference	---	5,072
Net Income and Taxes		
College	$144,560	---
High School	118,577	---
Difference	25,983	---
Federal, State, and Local Taxes*		
College	$21,925	---
High School	14,889	---
Difference	7,036	---

*Fiscal dividend is assumed to equal unity. Only that portion of local taxes received by state government included in the calculation; does not include those retained by local community.

The next item presented in Table 9 is the present values of the net income plus the total taxes paid by individuals over their lifetimes. These have been discounted at 5 percent, so it is possible to measure the incremental present value of college versus high school against other investment activities that the state itself might undertake. The present value of the college stream of lifetime income, including all taxes paid, at 5 percent is $144,560 and for high school is $118,577; the difference, in present values, of college versus high school at the 5 percent discount rate thus amounts to some $25,983. This increment, which arises solely because of a college education, is equivalent to the "state take." It measures the total incremental

---

opportunity income which has been forgone by not being a member of the labor force; thus the only cost elements to be charged against the $5,072 difference in our 8 percent example are such items as tuition.

contribution to gross state product which the individual makes because of his greater capital content. We have assumed in our calculations that the fiscal dividend, that portion of federal taxes paid which is returned to the state in the form of services, equals unity. The last set of figures in Table 9 shows the present value (at 5 percent discount rate) of the taxes paid by college graduates to federal, state, and local government units and the present value difference—which, for our figures, amounts to $7, 036.

A simple explanatory statement is needed here concerning the discrepancy between the sum of the present-value net income difference at 8 percent, the present-value tax difference at 5 percent, and the present-value net income plus taxes difference at 5 percent. Had we evaluated the net income for these occupations at 5 percent, then the difference between them, plus the difference between the tax streams, would sum to $25, 983 present-value net income plus tax difference.

What conclusions can be drawn from this analysis of income and tax streams? Table 10 presents the private rate of return to attending college, evaluated at 8 percent. This should be adjusted downward for the cost of attending college which would not have otherwise been incurred. In the case of California, where tuition is relatively low, this would amount only to about a $2, 000 difference. Next we present the social rate of return to attending college at 5 percent. This is simply the difference between the net income plus tax stream of the high school occupation and the college occupation. From this figure of $25, 983 we subtract present-value costs at 5 percent of producing a B. A. degree at the University of California. This amount, drawn from Table 8, is $9, 158. The remainder, a present-value sum of $16, 725, is what we refer to as the "net social return" at 5 percent.

We are able to conclude, in the case of the University of California and the state of California, that if alternative private investments yield 8 percent or less and if the cost of attending college to the individual is less than $5, 072, higher education is definitely a good private purchase.

If the alternative uses to which the state could devote the resources committed to providing current operating expenses at the University of California are earning 5 percent or less, then it can easily be seen that with a net social return of $16, 725 the purchase of undergraduate education by the taxpayers of California is definitely a good business deal. Even if we were to subtract from this $16, 725 the present value of the private cost to the individual who is securing his B. A. degree, we would still be left with a positive net social return at 5 percent which would be greatly in excess of present-value costs incurred.

Table 10 also shows the present value of the tax stream paid by college individuals over that paid by high school graduates. Arguments are sometimes made which justify the use of public funds to produce B. A. degrees on the basis that college graduates pay a larger total lifetime tax bill than high school graduates. However, after converting the differential tax stream to present value of producing a B. A. degree, we see that there is a tax earnings deficit for the production of a B. A. degree of some $2, 122.

This is equivalent to an annuity which, instead of yielding a positive sum, takes away the sum of $117. 90 each year for the forty-seven years between age eighteen and age sixty-five for each individual in whom the investment cost of $9, 158 for a B. A. has been made. Thus arguments for providing public resources for higher education which rest solely on the value of the incremental taxes paid by educated citizens are without foundation.

However, in the case of California, particularly the University of California, when one considers the total present-value increment to state product and state taxes secured by the investment of $9, 158 in the production of a B. A., the net social return figure of $16, 725 provides positive evidence of the fruitfulness of decisions to expend public resources for the production of B. A. degrees.

TABLE 10

Private and Social Returns to College Graduates

Private Return to College at 8 Percent*	$ 5, 072
Social Return to College at 5 Percent	$25, 983
Cost of a B. A. Degree	-9, 158
Net Social Return	$16, 725
Present-Value Tax	$ 7, 036
Cost of a B. A. Degree	-9, 158
Tax Earnings Deficit	-2, 122

*Less cost of attending college which would not otherwise be incurred.

A final recasting of the net social return concept makes crystal clear the very considerable benefits which can be realized from a system of public higher education such as that of California. For a state outlay of $9,158 (present value) at age eighteen for an individual, the state receives from each such individual who remains in the state of California for his working life an annuity for the entire working life of $1,444.55 per year.

SUMMARY

The purpose of this chapter was to investigate, using real income data on real-world occupations requiring college degrees and those requiring high school degrees, the rate of return secured by an individual and the rate of return secured by a state in the course of providing, at public expense, higher education which leads to a B.A. degree.

We have found, using such income data and actual current operating cost figures at the University of California for the production of a B.A. degree, that rather than being a marginal investment of public and private funds, investment in the production of human capital through publicly provided undergraduate education has an impressively high rate of return. Indeed, the substantial net social return points to the conclusion that rather than overallocating social resources for the provision of public undergraduate education, we may very well be engaging in substantial underallocation of resources in this socially productive area.

NOTES

1. For the original presentation by a modern economist of the concept of education as an increment to human capital, see T. Schultz, "Capital Formation by Education," *Journal of Political Economy* LXVIII (December, 1960); and "Investment in Human Capital," *American Economic Review* LI (March, 1961), 1-17. Later reviews are Burton Weisbrod, "The Valuation of Human Capital," *Journal of Political Economy*, LXIX (October, 1961), 425-36; and Gary Becker, "Investment in Human Capital: A Theoretical Analysis," *ibid*. LXX (October, 1962), 9-49.

2. For discussion of the problems of measuring public benefits or social benefits of public goods, see R. McKean, *Efficiency in Government Through Systems Analysis* (New York: Wiley, 1963); and Robert Dorfman, ed., *Measuring Benefits of Government Investments* (Washington, D.C.: Brookings Institution, 1966).

3. Gary S. Becker, Human Capital: A Theoretical and Empirical Analysis (New York: Columbia University Press, 1964); and Bruce Wilkinson, "Present Values of Lifetime Earnings for Different Occupations, "Journal of Political Economy, LXXIV (December, 1966), 556-72.

4. Bay Area Salary Survey Committee, The Joint Wage and Salary Survey (Sacremento: Department of Personnel, State of California, March, 1967); California State Personnel Board and United States Bureau of Labor Statistics, Cooperative Wage and Salary Survey (Sacremento: Department of Personnel, State of California, April, 1967). We wish to thank Mr. Lewis of the California State Personnel Board for his assistance to us in compiling the data for our specific needs.

5. S. Suslow, E. Langlois, R. Sumariwalla, and C. Walther, Student Performance and Attrition at the University of California, Berkeley (Berkeley: Office of Institutional Research, University of California, 1968).

6. For an elaboration of these see D. Bailey and T. Cargill, "The Military Draft and Future Income," Western Economic Journal, VII (September, 1969), 365-70.

7. For the use of present-value techniques in securing our optimal investment decision, including some of the pitfalls of their uncritical application, see Jack Wirshleifer, "On the Theory of Optimal Investment Decision," Journal of Political Economy, LXVI (August, 1958), 329-52.

ABOUT THE EDITOR AND THE CONTRIBUTORS

PAUL W. HAMELMAN is a member of the management sciences faculty at Virginia Polytechic Institute and State University, Blacksburg, Virginia. He has written over twenty papers and articles dealing with management science applications in colleges and universities which have appeared in numerous U.S. and foreign journals. His Ph.D. is from the University of Pittsburgh, where he was awarded the U.S. Steel Foundation fellowship in business administration. His professional affiliations include the American Association for the Advancement of Science, the Institute of Management Sciences, and the Operations Research Society of America (member).

DUNCAN BAILEY received his undergraduate degree from Stanford, his M.A. from the University of California, and the Ph.D. from Virginia Polytechnic Institute and State University. His articles have appeared in American Economic Review, Journal of the American Statistical Association, and Western Economic Journal. In addition, he has written or been coauthor of over ten papers on higher education planning. A principal in the firm of Crime Prevention Systems, Inc., his current interests include application of management techniques to improve the functioning of the criminal justice system.

DAVID C. CORDES is president of Management Decision Systems, Inc., St. Paul, Minnesota. He received his undergraduate degree in mechanical engineering, his master's in industrial engineering, and his Ph.D. in Production Management from the University of Minnesota School of Business Administration. He has worked in both the private and the public sectors: General Mills, the Office of Assistant Secretary of Defense (Systems Analysis), and as associate director of Project PRIME at the Minnesota Higher Education Coordinating Commission. His research papers and activities center on management systems (particularly PPB systems) and on management's utilization of the computer.

ROBERT H. FEARN is associate professor of economics and the institutional representative for the U.S. Department of Labor's Institutional Development Program at North Carolina State University, Raleigh. A native of New Jersey, Dr. Fearn received an M.A. from Washington State University and the Ph.D. from the University of Chicago. Prior to joining the faculty at NCSU, he was an economic intelligence officer with the Central Intelligence Agency. His publications include numerous articles and monographs dealing with labor, education, and housing problems in the Soviet Union and the United States.

LOREN A. IHNEN is professor and staff member of the Center for Occupational Education at North Carolina State University. He received an M.A. from the University of Illinois and a Ph.D. from Iowa State. His publications and research interests are in the area of agricultural economics, investment in human capital, and the economics of postsecondary educational systems.

BEN LAWRENCE is director of the National Center for Higher Educational Management Systems at WICHE, Boulder, Colorado. His undergraduate degree is from Whitman College, where he majored in mathematics and physics. His master's and Ph.D. degrees were earned at the University of Oregon. From 1966 to 1969 Dr. Lawrence served as executive director of the Education Coordinating Council for the state of Oregon.

WILLIAM E. MORAN is chancellor of the University of Michigan at Flint. After graduating from Princeton in 1954, he served as a line officer with the U.S. Navy. He resumed graduate studies after release from active duty and was awarded an M.B.A. by the Harvard Graduate School of Business and a Ph.D. from the University of Michigan. He has held positions with the firm of Booz, Allen and Hamilton; served on the administrative staff at Harvard; and was assistant executive vice-president of the State University of New York at Stony Brook immediately prior to being named to his present post. His contributions to the literature of educational planning and management have appeared in several professional journals.

CHARLES A. NELSON is a principal in the firm of Peat, Marwick, Mitchell & Co., New York. For the past fifteen years Mr. Nelson has been a management consultant for colleges and universities in the United States and abroad. Currently he and his colleagues are assisting several major universities in the design and installation of comprehensive MIS's. He is the author of a number of articles on educational planning and administration which have appeared in various professional journals.

CHARLES SCHOTTA is associate professor in the College of Business, Virginia Polytechnic Institute. He has written over fifty articles dealing with resource allocation aspects of colleges and universities, international economic development, statistical methodology, and monetary economics. His scholarly contributions have appeared in leading professional journals in the United States and abroad. Professional affiliations include the American Economic Association, American Finance Association, and the Operations Research Society of America (member).

MARTIN L. SHOTZBERGER is president of Catawba College, Salisbury, North Carolina. He is a graduate of the University of Richmond and holds a Ph.D. from Ohio State University. He has served as management development consultant to some of the nation's leading business organizations, as well as to governmental agencies.

JAMES F. TUCKER is professor of economics education, Virginia Polytechnic Institute and State University. He formerly served as president of Virginia State College and as director of the School of Business at North Carolina Central University and in the same capacity at West Virginia State College. He has served as full-time consultant to the West Virginia Board of Regents for Higher Education and is currently a consultant to the U.S. Office of Education. He has written a number of articles and monographs dealing with management and administration of educational institutions.